Dr. MICHAEL W. POPEJOY

The Way of
Duty, Honor, Country

American Warriors

Throughout the nation's history, numerous men and women of all ranks and branches of the United States military have served their country with honor and distinction. During times of war and peace, there are individuals whose exemplary achievements embody the highest standards of the U.S. armed forces. The aim of the American Warriors series is to examine the unique historical contributions of these individuals, whose legacies serve as enduring examples for soldiers and citizens alike. The series will promote a deeper and more comprehensive understanding of the U.S. armed forces.

Series editor: Roger Cirillo

An AUSA Book

The Way of Duty, Honor, Country

The Memoir of General Charles Pelot Summerall

CHARLES PELOT SUMMERALL

Edited and Annotated
by Timothy K. Nenninger

Original Manuscript and Research Assistance Provided by
the First Division Museum at Cantigny, Wheaton, Illinois

THE UNIVERSITY PRESS OF KENTUCKY

Scholarly publisher for the Commonwealth, serving Bellarmine University, Berea College, Centre College of Kentucky, Eastern Kentucky University, The Filson Historical Society, Georgetown College, Kentucky Historical Society, Kentucky State University, Morehead State University, Murray State University, Northern Kentucky University, Transylvania University, University of Kentucky, University of Louisville, and Western Kentucky University.

Editorial and Sales Offices: The University Press of Kentucky
663 South Limestone Street, Lexington, Kentucky 40508-4008
www.kentuckypress.com

14 13 12 11 10   5 4 3 2 1

Library of Congress Cataloging-in-Publication Data

Summerall, Charles Pelot, 1867-1955.
  The way of duty, honor, country : the memoir of general Charles Pelot Summerall / Charles Pelot Summerall / edited and annotated by Timothy K. Nenninger.
      p. cm. — (American warriors)
  Includes bibliographical references and index.
  ISBN 978-0-8131-2618-0 (hardcover : alk. paper)
  1. Summerall, Charles Pelot, 1867-1955. 2. Generals—United States—Biography. 3. United States. Army—Officers—Biography.
4. United States. Army—History—20th century. 5. United States—History, Military—20th century. 6. Citadel, the Military College of South Carolina—History. I. Nenninger, Timothy K. II. Title.
  E745.S86A3 2010
  355.0092—dc22
  [B]   2010030431

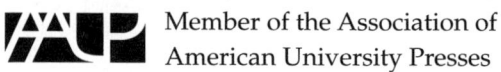

Contents

Photographs follow page 138

Editor's Note

About 1950, as an octogenarian, Charles P. Summerall recorded his memoir with a pencil on yellow foolscap, titling it "The Way of Duty, Honor, Country." According to his grandson, Dr. Charles P. Summerall III, of Charleston, South Carolina, the general wrote it largely "as a diversion without a plan to publish." After Summerall's death, his son typed the manuscript—single spaced, 185 pages. For many years, a copy was available to researchers at the archives of the Citadel, where Summerall had been superintendent from 1931 to 1953, and where a collection of his papers is housed. In 1998, the Cantigny First Division Museum received the rights to the manuscript from Charles P. Summerall III and enlisted my services to prepare the memoir for publication.

Working from a copy four generations removed from the original, I have endeavored to provide the sort of copyediting that a publisher would have required had Summerall sent his draft manuscript to a commercial, academic, or military specialty outlet at the time he wrote it. Among other work, three-page paragraphs have been reduced to more manageable proportions, while successive single-sentence paragraphs have been combined to produce more readable copy. The overall effort has been to improve syntax, readability, and narrative flow. Summerall had an important and interesting military career, and his memoir reflects that. But the manuscript, written many years removed from the events described, often rambled and included tangential observations unrelated to the main flow of the narrative. I have attempted to bring focus to the effort that, on occasion, has resulted in the deletion of perceived irrelevant or misplaced passages.

The simple copyediting in particular attempted to correct spelling, punctuation, and grammatical errors as well as standardize capitalization. A significant effort involved the identification in notes of people, places, things, and events not clearly or fully described in the text. In a few cases, mostly involving people, additional identification proved impossible, and Summerall's identification, or lack thereof, remains unchanged. Also, I did not always include annotations for well-known individuals or those for whom Summerall provided description or context.

A variety of sources, published, manuscript, and human, assisted in this effort. Published sources included dictionaries, geographic dictionaries, gazetteers, atlases, registers of U.S. Army officers, West Point and Annapolis alumni registers, *Who's Who,* and biographical and military dictionaries as well as a range of histories, memoirs, and monographs that concern the army from the late nineteenth century to the mid-twentieth. The occasional Google search in a number of instances solved difficult conundrums of identification, particularly of obscure individuals mentioned in the text. Military records in the National Archives were frequently the crucial source for much of the identification and annotation effort, as were the forty-three boxes of the Charles P. Summerall Papers in the Manuscript Division at the Library of Congress. The result is over five hundred annotations, identifications, and other notes. Most frequently, the annotations provide brief biographical information about individuals. Because the various elements of the identification (birth, death, and retirement dates, military rank, important positions held, etc.) for a particular individual frequently came from multiple sources and often from different annual editions of the same publication (different years of the *Army Register, Who's Who,* or the West Point *Register of Graduates,* e.g.), I usually did not provide a specific citation to each published source consulted for this information. In instances where the information came from unpublished manuscript or archival material, source citations are included. General Summerall mentions many more than a hundred individuals in the text, including army colleagues, relatives, family friends, and others with whom he had contact. I have attempted to provide some basic biographical information about as many of these people as possible. The military identifications were usually possible to determine, but, in some instances, it was not possible from manuscript or other sources to provide positive identification or complete biographical information.

Richard Peuser, Trevor Plante, and Mitchell Yockelson at the National Archives assisted in identifying relevant records and needed files. Robert Cressman at the Naval Historical Center and John J. Rhodes at the Marine Corps History Division provided information about ships and people. Jane Yates at the Citadel archives looked up needed biographical information about several of Summerall's associates at that institution. Holly Reed at the National Archives, Tim Frank, an independent researcher, and Andrew Woods at the First Division Museum provided crucial support in locating and copying relevant photographs.

My friends and professional colleagues the late Larry Bland, Edward M. Coffman, and Brian Linn provided substantive information when I was fact-checking but also offered useful more general advice on how to edit the manuscript. John Votaw conceived of the need to publish Summerall's manuscript, and he and Paul Herbert, his successor as the director of the Cantigny First Division Museum, supported the project throughout. John Lindley helped with the early copyediting, while Joseph Brown did the final polishing. Many thanks go to Roger Cirillo, the series editor for Association of the United States Army Books, who late in the game expressed interest in publishing the manuscript and encouraged me to complete the project. Stephen Wrinn at the University Press of Kentucky ably oversaw the final stages of the process and brought the manuscript to publication. Finally, Dr. Charles P. Summerall III, the general's grandson, donated the manuscript to the Cantigny First Division Museum and encouraged its publication.

Chronology

March 4, 1867	Born, Blount's Ferry, FL
1882–1885	Attended Porter Military Academy, Charleston, SC
1885–1888	Taught school in Florida
1888–1892	Attended U.S. Military Academy, West Point, NY
1892–1893	Second lieutenant, First Infantry, Benicia Barracks, CA
1893–1896	Fifth Artillery, Presidio of San Francisco, CA
1896–1898	Fort Hamilton, New York Harbor
1898	Aide-de-camp to commanding general, Department of the Gulf
1899–1900	First lieutenant, Fifth Artillery, in action against insurgents in the Philippine Islands
1900–1901	In action with the China Relief Expedition, Peking
1901	Marries Laura Mordecai, daughter of Brigadier General Alfred Mordecai, Ordnance Department
1901–1902	Captain, 106th Company Coast Artillery, Fort Walla Walla, WA
1901–1903	Post commander at Fort Lawton, WA; Camp Skagway, AK; Fort William H. Seward, AK; and Fort Flagler, WA
1903–1905	Third Battery Field Artillery at Camp George H. Thomas, GA, and Fort Myer, VA
1905–1911	Senior instructor of artillery tactics, U.S. Military Academy, West Point, NY

1907	Assigned to Third Artillery, then transferred to the Second Field Artillery
1911	Promoted to major, Third Field Artillery, and given command of a field artillery battalion with the Maneuver Division, San Antonio, TX
1911–1914	Commanding battalion with Third Field Artillery at Fort Myer, VA
1914–1917	Assistant to chief of the Militia Bureau and responsible for national guard field artillery training
1916–1917	Promoted to lieutenant colonel, July 1, 1916; promoted to colonel, May 15, 1917; promoted to brigadier general, August 5, 1917
1917	Member of the Baker Mission, a military mission to investigate the organization of British and French armies in Europe
1917–1918	Commanding officer, Sixty-seventh Field Artillery Brigade, Forty-second Division (September 5–December 22, 1917); then commanding officer, First Field Artillery Brigade, First Division (December 23, 1917–July 17, 1918)
1918	Promoted to major general, June 26, 1918
1918–1919	Commanding general, First Division (July 15–October 12, 1918); then commanding general, Fifth Corps (October 12, 1918–February 12, 1919)
1919	Commanding general, Ninth Corps (February 28–April 16, 1919); commanding general, Fourth Corps (May 2–June 22, 1919); member, Inter-Allied Commission of Inquiry on Fiume, Italy (July 8–August 14, 1919); member, American Commission to Negotiate Peace (August 14–31, 1919)
1919–1921	Commanding general, First Division (September 30, 1919–June 30, 1921), at

	Camp Dix, NJ, and Camp Zachary Taylor, KY
1921–1924	Commanding general, Hawaiian Department (August 5, 1921–October 5, 1924)
1924–1925	Commanding general, Eighth Corps Area, Fort Sam Houston, TX (October 5, 1924–January 16, 1925)
1925–1926	Commanding general, Second Corps Area, Governor's Island, NY (January 16, 1925–November 20, 1926)
1926–1930	Chief of staff, U.S. Army (November 21, 1926–November 20, 1930)
1929	Promoted to general, U.S. Army, February 23, 1929
1931	Retired, March 31, 1931
1931–1953	President, the Citadel, Charleston, SC (retired June 30, 1953)
May 14, 1955	Died at Walter Reed Army Hospital, Washington, DC

Introduction

Charles P. Summerall in History

Charles P. Summerall, a son of the poor rural South from the post–
Civil War era, entered the army before the last of the major battles
with the Plains Indians, served on active duty as the army adjusted
to its growing role as guardian of American insular possessions, and
was an important combat commander when the army first entered
the global stage as a participant in a European war. His active duty
concluded with him at the pinnacle of his profession, leading the
army as chief of staff, the senior uniformed officer. Thus, by a number of measures, Summerall had a more than successful military
career during a period of profound change for the service. After retirement, he began what became a long and equally successful second career as a college educator—president of the Citadel.[1]

Summerall's military leadership qualities were recognized early
on. In his second, third, and fourth years at the U.S. Military Academy, he served, respectively, as cadet corporal, first sergeant, and first
captain of the Corps of Cadets. After commissioning, his service as a
junior officer was similarly distinguished, with citations for bravery
under fire and recognition as an administrator, trainer, and leader.
While commanding an artillery battery in a firefight against Filipino
insurrectionists, on January 1, 1900, Summerall twice pressed his
guns forward to bring the enemy under direct fire and, thus, relieve pressure on the infantry his artillery was supporting. His example under heavy enemy fire "plainly and unmistakenly inspired"
all American troops in view.[2] Eight months later, in the attack on
Peking during the China Relief Expedition, he again demonstrated personal bravery and professional skill in utilizing his battery's
guns to blow open four successive gates leading into the Imperial
City. During the prewar years, his superiors rated Summerall "an
artillery officer of exceptional merit" and "one of the best light artillery officers in the service" and the battery he commanded "the best
I have ever seen."[3]

In 1912, Summerall went to the War Department Militia Bureau to work on issues relating to the organization and training of state militia field artillery units. Among other accomplishments, he surveyed and negotiated leases for sites in the Eastern United States to be used as artillery firing ranges. He supervised the development of a site at Tobyhanna, Pennsylvania, that became the major East Coast field artillery training center during World War I. For four years (1912–1914, 1916), he was an instructor in artillery tactics at joint army–national guard (militia) summer training camps. His prowess with technical aspects of artillery was developed as a member of the War Department Ordnance Board (1915–1916) and the Board of Ordnance and Fortification (1917).

Five weeks after the United States entered the First World War, the War Department promoted Summerall to colonel and sent him to France as the artillery expert with a mission of American officers, headed by Colonel Chauncey Baker, that would make recommendations on proper organization and doctrine for the American Expeditionary Forces (AEF) then forming overseas. Summerall concluded that, in order to achieve decisive results when attacking, American forces had to use artillery "in quantities not hitherto contemplated" and had to perfect artillery tactics, preferably using rolling barrages. Although his recommendations, for 256 guns for every thousand yards of front, did not conform with what General Pershing's headquarters eventually accepted, 118 for every thousand yards of front, his work with the Baker Mission during June and July 1917 further enhanced his reputation as a field artillery organizer and tactician.[4]

On return to the United States, Summerall received a promotion to brigadier general and command of the Sixty-seventh Field Artillery Brigade, Forty-second Division, which he took overseas for its initial training in France. But soon after arrival, in December 1917, he received command of the First Field Artillery Brigade, First Division, and his service with that organization has led some historians to conclude that he was the best American artilleryman of the era.[5] Among other innovations, his batteries perfected the creeping barrage to precede the advancing infantry and the use of direct fire by guns located with frontline troops. In July 1918, General Pershing, recognizing Summerall's achievements, appointed him to command the First Division, replacing General Robert L. Bullard, who had been promoted to a corps command.

Summerall led the First Division in three major operations—Soissons, St. Mihiel, and the first two weeks of the Meuse-Argonne.

At Soissons, despite heavy casualties and sporadic support from the French, the First Division achieved all its assigned objectives. In the attack, Summerall, who had taken command only the day before the operation commenced, made a reconnaissance under fire, personally exhorted his men to renew an assault on the enemy position at Berzy-le-Sec, and generally inspired his troops by his example. For his gallantry, he received the Distinguished Service Cross. By most accounts, he skillfully handled the division at St. Mihiel and in the opening phase of the Meuse-Argonne. For the rest of his life, his name was nearly synonymous with the First Division—as a distinguished combat commander of the division artillery brigade and the division (December 1917–October 1918), for his command after the division returned to the United States (September 1919–June 1922), and for his active role in forming the division veterans' association, publishing its wartime history, establishing the division war memorial in Washington, DC, and generally promoting and perpetuating the reputation of the First Division.

As the Meuse-Argonne attack stalled in mid-October 1918, Pershing made changes in the First Army leadership, relieving several division and corps commanders. He selected Summerall to lead the Fifth Corps during what became the final stage of the war. On November 1, the offensive resumed with the Fifth Corps leading the attack in the First Army center. American infantry and artillery finally broke through the main German defenses. The pursuit following this breakthrough led to competition among several Allied units to liberate the French city of Sedan. The "race to Sedan" tarnished Summerall's largely unblemished World War I reputation and poisoned his relations with a few other senior AEF commanders and general staff offices, notably Major General Joseph T. Dickman, who commanded the adjacent First Corps. The First Army order detailing the advance on Sedan was not well written or especially clear as to responsibility for seizing the town. Summerall used this ambiguity to justify ordering the First Division to move out of its zone of operations in the Fifth Corps area and across the advance of the Forty-second Division in the area of the First Corps on the left. Considerable chaos and confusion resulted. Summerall was not alone in the responsibility for this debacle. But he put his personal feelings, wanting his corps and the First Division in particular to beat other units to Sedan, above his professional judgment. With the imminent collapse of the German army and the subsequent Armistice, the Sedan incident had no lasting military impact. But it caused continuing animosity long after the war among several senior AEF general officers; Summerall's long-lasting enmity toward several colleagues, among them Joseph T. Dickman and Malin Craig (commander and chief of staff of First Corps, respectively), stems in large part from Sedan.[6]

Despite Sedan, as a combat commander—brigade, division, and corps—Summerall rivaled any American general in terms of service in the AEF. Pershing's selection of him to command the First Division—probably the best AEF formation and Pershing's favorite—and the Fifth Corps indicated that the commander in chief thought highly of him.[7] Initially, Hunter Liggett, the First Army commander during the final weeks of the war, considered Summerall "in a class by himself." But Sedan caused Liggett to temper his enthusiasm.[8] Some doughboys who served under Summerall had a more negative view, regarding him as a ruthless driver regardless of the cost in lives. As Father Francis P. Duffy, the chaplain of the Forty-second Division, put it: "He wanted results, no matter how many men were killed."[9] But after the Armistice, and in the afterglow of apparent American military success, Summerall's role in the Sedan incident and his sanguinary reputation among his troops were forgotten. He was remembered as an organizer, an innovative tactician, and a successful combat commander, one of a handful that had led a corps in battle.

Following the Armistice, Summerall remained in Europe until September 1919, commanding, successively, the Fifth, Ninth, and Fourth Corps in the Occupation, and serving with the American Commission to Negotiate Peace. Notably in the latter role, he was the American representative on the Inter-Allied Commission of Inquiry on Fiume, Italy, which sought to resolve boundary disputes in the region. He returned to the United States once again in command of the First Division. His subsequent postwar assignments were appropriate to his rank and reputation, including command of the Hawaiian Department, the Eighth Corps Area, headquartered at Fort Sam Houston, Texas, and the Second Corps Area, with headquarters in New York.

On September 21, 1926, the War Department announced that Summerall would succeed Major General John L. Hines as chief of staff of the army. This was a logical choice as Summerall was the senior major general of the line in the army and virtually no other general had a record comparable to his. When he assumed his duties on November 21, his post and the army he was to lead had both been shaped by outside forces that he neither completely liked nor could significantly change. By the end of 1926, the army had probably reached the nadir of its interwar existence. That year saw the low point for War Department expenditures and nearly the low point for overall army strength. During this period, the army may have been less ready to function as a fighting organization than at any time in its history. At best: "As anything more than a small school for soldiers the Army scarcely existed."[10]

Charles P. Summerall will not be remembered as a great chief of staff of the army. His personality, his immediate professional interests, and events during the time he served combined to limit his opportunities and his ability to achieve greatness. There was no national emergency or even much of a creditable threat to American security during his time as chief. Although the War Department was the second largest government agency of the period, most of its activities were basically civil functions and the responsibility of the department's civilian leadership. By temperament, Summerall was little concerned with matters of national policy or even military strategy. He did serve on the Joint Army-Navy Board, recognized the importance of joint war planning, and was particularly concerned about the reinforcement of American insular garrisons in the event of war. But internal army problems, those that he could most directly affect, were of greater concern, and they dominated his incumbency as chief of staff. Summerall was an army traditionalist to whom soldier morale was important, although he could be hard on his troops and subordinates. Similarly, while he lent important support to mechanization, he was much more skeptical about the growth of military aviation because it was coming largely at the expense of traditional arms. Apart from personal feuds with some of his senior officers, his biggest battles as chief were over the army budget. From 1926 to 1930, given concern about economy in government during the Coolidge administration and the onset of the Depression during the Hoover years, significant increases in army spending were impossible. But in several key areas Summerall was able to influence events and gain some modest increases—specifically for housing, the daily ration, and mechanization. His overseeing of a 1930 survey of the military establishment was particularly skillful, ultimately convincing Hoover that, far from requiring cuts, the army's budget should be stabilized to sustain essential services. At heart, Summerall was a conservative chief for a conservative era—holding the line on budgets, making incremental improvements in internal army conditions, but breaking little new ground.[11]

Summerall relinquished his duties as chief of staff on November 20, 1930, and retired from the army on March 31, 1931. He briefly considered an attempt at a political career but eventually eschewed politics for education. In September 1931, he accepted an appointment as the president of the Citadel. Under his leadership for the next two decades, the school doubled its enrollment, balanced its budget, improved its faculty, and secured university accreditation. It grew from a small Southern military school to a national institution.[12] In 1953, at age eighty-six, Summerall retired from the Citadel to live in Akin, South Carolina. The next year, a terminal illness forced him to Walter Reed Army Hospital in Washington, DC. On May 14, 1955, he died in the hospital and was buried in Arlington National Cemetery.[13]

Charles P. Summerall lived an interesting life filled with accomplishments. His military service came during a period of rapid and fundamental change for the U.S. Army. For these reasons alone, his memoir is worth reading. But he was also a man of strong convictions and decided opinions. He used the memoir to reiterate them, often to validate his actions, and to make clear who his friends and perceived enemies were and why. These characteristics all make *The Way of Duty, Honor, Country* an interesting read.

Chapter 1

The Rock Whence
I Was Hewn

I have no records of the original Summeralls in America. By family tradition, it is known that Thomas Summerall went to Florida and married a Spanish woman named Neunez in St. Augustine. He lived in south Florida by raising cattle and shipping them to Cuba. A son of Thomas Summerall named William was born probably about 1805 in Wayne County, Georgia. He married Hetty Wiggins and settled at Blount's Ferry on the Suwannee River in Columbia County, Florida. Here, he acquired a plantation, owned slaves, and kept a store. When the Civil War came, he was quite prosperous. He had a number of children, including at least three sons and three daughters. The eldest, Elbanan Bryant Summerall, born July 5, 1827, remained on the plantation until he went into the Confederate army. In 1862, he married Margaret Cornelia Pelot at Greenwood, South Carolina. They were my parents. My sister, Meta Margaret Ann Summerall, was born July 4, 1863, my brother, William Bryant Summerall, was born April 29, 1865, and I was born March 4, 1867, all at Blount's Ferry. During the Civil War, my father served in a Florida regiment[1] that took part in opposing the Seymour expedition[2] in Florida. He was mustered out on account of sickness. Throughout his life, my father had a passionate desire to study medicine, and his whole heart was in medical subjects. This aim was thwarted by the war, and his life was frustrated in consequence. He devoted all the time he could find to reading medicine, and, although the family had many illnesses, a doctor was rarely called. He treated us as effectively as the doctor could have done. We were too poor to pay doctors if we had needed them.

After the war, when the slaves left the plantation, my grandfather moved to near Lake Butler, Florida, where he tried to farm and raise cattle. Later, he again moved to Falling Creek near Lake City, Florida, where he raised cattle and hogs on a small farm. For what reason I never knew, my grandfather and his first wife, Hetty Higgins, were divorced, and he married another woman by whom he had several children. My grandmother Summerall lived with us for several years at Providence, Florida.

My mother's home was Greenwood, South Carolina. She taught school from girlhood in various towns in South Carolina and in the Columbia Female Seminary. She also taught school in Quincy, Florida, where she probably met my father. For several years, my mother taught school at Providence, Florida, and in neighboring places. The pay was small, and the term was only three months. The people were impoverished by the war, and the struggle for a living was desperate.

The records of my mother's genealogy are quite complete. Jean Pelot and his wife, Marie Bossant, were French Huguenots who went to Switzerland on the Edict of Nantes.[3] Their son Jonas Pelot (born 1695, died 1768) and his wife, Suzanne Marie Paquet, emigrated to South Carolina in the early migration. They probably settled in Purysburg on the Savannah River. They had three sons: John Francis Pelot, born in 1720, died in 1774; Captain James Charles Pelot, born in 1763, died in 1809; and Charles Moore Pelot, born 1791, died 1863. He [Charles Moore] married Margaret Ann Ford, and their daughter, Margaret Cornelia Pelot, married Elbanan Bryant Summerall.

Charles Moore Pelot and his family left Purysburg about 1832, when it was abandoned,[4] and settled at Greenwood, South Carolina. At the time of his death, Pelot appeared to live in Cokesbury, South Carolina. All the rest of the Pelot family left South Carolina and settled in Missouri immediately after the Civil War. My grandfather was an intellectual man without success financially. He taught school and was the railroad agent and postmaster at Hodges, South Carolina, at the time of his death.

My mother and her brothers and sisters had unusual talent and character. Her brother James Malachi Pelot graduated from the Citadel, Charleston, South Carolina, studied medicine, and was a surgeon in the Confederate army. Thomas Postell Pelot, another brother, graduated from the U.S. Naval Academy. When the Civil War began, he resigned from the U.S. Navy and was commissioned in the Confederate navy. He had a crew at Savannah but no ship. On a stormy night, he took his crew in rowboats and boarded and captured the U.S. cruiser *Waterwitch*. In the hand-to-hand fighting, on deck, he was killed. It was a daring and gallant deed and is recorded in the official records of the Union and Confederate navies.[5] One sister, Caroline, married Dr. Davison in Missouri. Their son Gregory C. Davison made a brilliant record in the navy.[6] Another sister, Julia, married Mr. R. P. Motte of the distinguished Huguenot family in South Carolina.

Chapter 2

The Pit Whence
I Was Digged

When I was born in 1867, the South was in the throes of Reconstruction and what has aptly been called the Tragic Era. Without the slaves, the plantations were useless. The most extreme poverty and widespread suffering prevailed. On leaving Blount's Ferry about 1870, my father moved to Providence, Florida, about sixteen miles from Lake City, where he tried to operate a wheelwright shop. For several years, my mother taught there in the country schools. The pay was small, and the term was three months. She boarded around among the patrons, walking three miles over the country paths, often in the rain. How she bore it, frail as she always was, can only be explained by her indomitable spirit and brave fortitude. She was always cheerful and never complained. The war had impoverished the people, and their struggle to live was desperate. Because the land was poor, they depended entirely on the food they could raise and the cotton and poultry they could sell at very low prices. They could pay my father for his work only in produce, which was not sufficient to support the family. Our food was meager, consisting mostly of our garden vegetables, potatoes, corn and corn bread, hominy, and very little meat. It was necessary to sell chickens and eggs to buy cloth for the clothes that my mother made for us. The boys did not have shoes. My mother made our hats from palmetto,[1] which she plaited with our help. Our house was a small, one-room, very old log cabin. There were times when little or no food was in the house.

From my earliest recollection, I thinned cotton or planted potato vines or picked cotton. As I grew older, I hoed and plowed at ten or eleven. I was proud to do the work of the men. We played little except during the time we were at school. We had few toys. I was the baby and was treated as such till I was six or seven. My mother kept my blond hair down to my shoulders and curled it every day. The boys teased me and called me "Sally," much to my annoyance. We all sang, and I was called on as a small boy to lead the singing in the occasional religious services. Once, I was routed out of bed at night to lead the singing at a church service. There was a small building belonging to the Masons, called the lodge. It was used for every purpose. Occasionally, an itinerant preacher conducted a service of a primitive kind on Sunday. Now and then, we walked, for we had no horse or mule, about three miles to an occasional Baptist service at a church on Olustee Creek. Because my mother taught us every Sunday, we knew the Bible well.

In 1879, an old friend of my mother's, Mrs. E. R. Moore, came to Providence from South Carolina to teach the local school. She told my mother about the Reverend A. Toomer Porter and the school that he had established in Charleston, South Carolina, for poor boys, called the Holy Communion Church Institute.[2] My mother wrote to Dr. Porter and asked him to take my brother. This he did without charge at the school. The railroads gave passes to Dr. Porter's boys, and in some way my parents found a small sum to pay for his uniform and a suit of clothes. He was very bright and did well.

The country around Providence was dominated by a lawless family and their gangster associates, who committed many murders. My father, one of the small number of good citizens, incurred the enmity of the outlaws by being appointed justice of the peace and postmaster. They made two attempts on his life, firing at him in the dark. He always carried a shotgun loaded with buckshot. When the conditions became intolerable, he sold the small place at Providence and moved to Live Oak, Florida. There, my mother taught at another three months' school, but my father could not find work. He collaborated with a doctor in trying to manufacture a sore eye medicine but was unsuccessful. As a result, in 1880 he went to Umatilla, Florida, to make a start in the migration to this new country. My mother and I remained at Live Oak, where we boarded with a fine farm family named Kennedy. My sister also went to Umatilla to teach. My mother and I joined my father and sister at Umatilla in 1881, where my mother again taught school. My sister was able to attend the Columbia [South Carolina] Female Seminary with the money she had made teaching. In 1882, my father went to Eustis, Florida, and tried to merchandise, without success.

That year, Dr. Porter accepted me as well as my brother, and a new world opened to me. Whatever I have done in life is due to this beginning. I stood well in my class at Dr. Porter's and generally took a leading part. As monitor of a dormitory, I was responsible for order and compliance with the rules. I was active in the school literary societies and sang in the church choir. I learned to dance and went out with young girls when there were occasions and no expense was involved. I had no money. I was baptized and confirmed in the Holy Communion Church. I spent my vacation in 1883 at Eustis. As monitor of the choir, I had one serious difficulty. A boy was disorderly during a very solemn service in the chapel. When I told him to behave, he said a very vile thing to me so others could hear. As soon as the choir marched into the vestry room, I demanded an apology, to which he sneered. I knocked him down in the presence of Dr. Porter, who was furious, ordering me to leave the school. But, when I explained, he said no more about it. He quite sympathized with me, though he did not punish the other boy. It taught me to control my temper. I should have called him out and fought him. He was my own size and age. Dr. Porter's school occupied the old U.S. arsenal, which the federal government gave him for the purpose. He built the four-story dormitory and the chapel. There were about 150 boarders and as many day students of ages from nine to over twenty. We wore civilian clothes except when we went into the city, and then we wore the uniform. I had only one suit and never thought of an overcoat till the last year, when my brother, who had been employed as a teacher, gave me one. He also paid the charge of the school of $210.00 for the last term. In after years, I returned this to him many times over and paid Dr. Porter for my expenses the first two years.

In 1884, my mother procured a school at Astatula, Florida, where my parents moved. In 1885, my mother taught at Minneola, Florida, where I went on graduating from the Holy Communion Institute in June 1885. That fall, my parents moved back to Astatula, and my father opened a small store, which I tried to keep, but the patronage did not support it. There were no openings for employment to be found. We cultivated orange groves for absent owners and painted houses in Mount Dora and Apopka City, making very little. We lived in an old, dilapidated log house with two small piazza rooms and a kitchen in the yard where we cooked and ate. I helped my mother with the cooking and did much of the washing at the lake.

In the winter of 1886, I taught at the Astatula School for three months at $30.00 per month. All this was used for our living. My father and I continued to find painting jobs a part of the time. In the winter of 1887, I taught school in Leesburg, Florida, for three months at $60.00 per month, but I paid rent of $25.00 per month for the school building. During the Christmas holidays, I passed a special examination for a state certificate, and my pay was increased to $85.00 per month. This enabled my father to buy a lot and lumber and build a house at Astatula. I boarded at the Commercial Hotel in Leesburg at a very small cost. I was also able to support my parents.

When Grover Cleveland was elected president, there was great rejoicing in the South over a Democratic victory.[3] The people took heart and thought that a new day had dawned from the night of the Civil War. I had the temerity to write to Mr. Cleveland to ask for an appointment to the Naval Academy. Of course no attention was paid to my request. But, during the 1887 Christmas holidays, I read in the *Leesburg Commercial* that the local congressman[4] would hold a competitive examination for appointment to West Point in Jacksonville. I took a train to Jacksonville the morning of the examination. It was held at a schoolhouse, where more than a dozen young men competed. A few days later, I was announced as the winner. The congressman did not nominate me, however, and made no reply to letters from my mother and me. At that time, all candidates reported to West Point about June 11 to be examined for admittance. The time was approaching, and it became evident the congressman did not intend to appoint me. He had appointed a straw board for the examination consisting of Mr. Sherman Conant, the great Florida railroad builder; Mr. C. H. Jones, the editor of the *Jacksonville Times Union*; and the superintendent of the Jacksonville schools. I happened to meet Mr. Conant in the hotel in Leesburg toward the end of May, and I told him of my predicament. He was furious and said, "That man cannot treat you that way." In a few days, I had orders from the War Department to report at West Point. The editor of the paper in Eustis, Mr. Hill, procured a pass for me on the steamer from Jacksonville to New York. I had little money left because of the expense at home.

Chapter 3

"And David was wise in all his ways and the Lord was with David"

On reaching New York, I went to the dock of the riverboat *Mary Powell,* where the agent asked me if I wanted a return ticket. I told him no. Opportunity had opened up to me, and I could not fail. On June 11, 1888, I reported to Lieutenant W. C. Brown,[1] the U.S. Military Academy adjutant, and was sent to Captain Spurgin,[2] the quartermaster, to pay the deposit of $65.00 required of candidates. I had only about $20.00, and I told him that I would pay the rest later. He was very angry but accepted what I had. I do not recall how my parents obtained the balance. The ordeal of Beast Barracks[3] began at once. I found myself in a room in the old Ninth Division with a man from the Ozarks and one from Vermont. Both were typical of their sections, as I, no doubt, was typical of mine. The rough treatment and harsh manner of the cadets over us was a great shock, but because all were undergoing the same experience, we tried to comply with our surroundings.

The examinations lasted about a week. Something like half of the candidates were rejected. I was devoutly thankful that I had passed, for I could not think of failure. It meant hope and life. I hated to leave my parents to make their own living, but a kind Providence sent a very fine man, Mr. McGuire, who owned a grove, to board with them. What he paid them, with a few other boarders at times and the work my father could find, enabled them to live. I never received a cent from anyone while I was at West Point and never bought a class ring. I did not even have my photograph taken as first captain.

I was overawed by the magnitude of the place; the superb appearance of the officers, who seemed to be supermen; the dashing bearing of the cadet officers; the superiority of all cadets; the comforts of living with such clothing, rooms, and food as I had never known or imagined; and the exhausting drills and exercises that filled the time. Horrible as was the treatment, I wanted to stay and be a part of it with all my being. Neither there nor elsewhere did it ever occur to me to be homesick. At twenty-one, the hardships and poverty I had known and the promise of a career instead of hopelessness made me accept whatever came in a philosophical way. I respected, admired, and revered the place and was proud and thankful to be a part of it. I did not then realize that the process would completely change my mental development by changing all that I had been to what it was intended that I should become. No doubt, the change was for the better in every way.

The chaplain's text for the first sermon preached after we entered was "Look unto the rock whence ye were hewn and to the hole of the pit whence ye were digged."[4] I felt that it had an especial meaning to me. The text of the next sermon was "And David was wise in all his ways and the Lord was with David." Again, I took it to heart and hoped and prayed that I might be wise and that the Lord would be with me. Among the many fortunate events that attended me were my assignment to "A" Company and my being in a tent in plebe camp with Leonard M. Prince[5] and John McA. Palmer.[6] They were recognized at once as being among the leaders of the class. Prince was inclined to be fresh, or "B.J.,"[7] and attracted much attention from the yearlings.[8] He became one of the leading athletes and, with Dennis Michie,[9] was a father of football at the academy. Palmer early displayed the superior intellect for which his career as a cadet and as an officer was noted. He wrote class songs and verses of a high order, and his contribution to history and literature in the army has not been excelled.[10]

We took our ordeal and licked our wounds with stoicism. I had never learned to swim. We were taught to swim in the Hudson River, where a small float was located near the rifle range in Washington Valley. A rope was tied around us, and in we jumped. I could not make the strokes, and after struggling as long as I could bear it without drowning, I was pulled out. This continued for weeks. One day, after I watched a cadet make the strokes, I jumped in and found no difficulty in moving my hands and legs as he did. We were required to swim ten minutes to qualify. The other cadets told me to continue till at last they told me to come out. I had swum sixty-three minutes with perfect ease. I never went to Washington Valley again.

I had no difficulty learning the manual of arms and infantry drill, but the artillery drill was very complicated. We were taught at a field battery of muzzle-loading guns, which were the type used in the army since the Civil War. The duties of No. 1 [member of the gun crew] required a number of movements to sponge the gun, ram the charge, and then ram the projectile. At my first drill, I was No. 1. A cadet private from the first class was in charge of the gun. He abused me and vilified me at the top of his voice, calling me all that was vile, and denouncing me for every offense. It was a most cowardly and dastardly abuse of authority. He reported me for a long list of offenses, which I could not have committed. When this was published at parade, I felt keenly the injustice and humiliation. I went to the commandant of cadets the next morning and told him of my utter ignorance and my efforts. He promptly removed the reports. Later, that cadet was placed in light prison for insubordination. He became a worthless officer, retiring without rendering any worthwhile service. In after years, I, as a captain of a light battery, had a collision with him when he tried to interfere with my watering the horses in maneuvers.[11]

I was fortunate in being able to sing and in knowing some French and German songs, which I had learned at Dr. Porter's. The upperclassmen made me go to the color line in the evenings and sing "La Marseillaise," "Die Wacht am Rhein," and other songs for the amusement of their young lady friends.[12] We were taught dancing with one another.

We learned the important subject of guard duty from Cadet Young,[13] a Negro first classman. He commanded the guard, marched us to meals, sat at the head of the table, inspected reliefs and sentinels, received salutes, and was treated with every official courtesy. No one felt any resentment or objected, and I accepted the situation as a matter of course.

At the end of plebe camp, I was again fortunate in having Cadet Donald W. Kellogg[14] ask me to room with him in barracks. He had entered the year before and failed. He was turned back and joined my class. He was a man of unusual force and character and had many friends in the yearling class. His association with me protected me from harassing by the upperclassmen, although hazing was supposed to stop in barracks. We were able to select the left plain room on the first floor of the first division, opposite the suite of the cadet adjutant. Kellogg's experience and knowledge of life in bar-

racks were a great help to me, especially in keeping our room in superior condition. Unfortunately, he could not learn the mathematics
as taught and was discharged for failure at the end of the year. My
friendship with him never ceased, though I never saw him again.

We were assigned to sections in alphabetical order for the first
few weeks, after which we were reassigned according to the grades
that we made. I was in the first or second sections in English and
French and the third section in mathematics. This was due to the excellent teaching at Dr. Porter's and to my teaching experience. The
lessons were long and recitations difficult. Seemingly, the text had
little to do with what we were expected to know, and the instructors
tried to find out what we did not know and mark us accordingly.
We refrained from asking questions because of the belief that the
instructor would mark us low for what we did not know. But the
officers who taught us commanded great respect and forced us to
learn for ourselves instead of depending on them. We also gained
much information from other cadets of the class.

Guard duty and drills continued, and life was very intense. I
devoted practically all of my time to study. There were no athletics except one hour every day in the gymnasium. We braced[15] in
marching to meals and in the mess hall after eating. I experienced
my first real cold and saw snow for the first time that winter. In the
dead of winter, a military funeral was held for General MacKenzie.[16] We stood on the plain in snow and ice with a cutting north
wind while the service took place in the old chapel. The minute
guns, the caparisoned[17] horse led after the caisson, the funeral
march by the band, and the Corps of Cadets with arms reversed
at common time brought to my mind the beautiful verse in the
"Burial of Moses."[18]

More and more, the military life captivated me, and I wanted
with all my heart to become a part of it. The parades and guard
mounting, the program from reveille to taps, and the precision everywhere ennobled the life of the soldier. I was tired at night and
slept soundly when we were allowed to go to bed, except when a
lesson forced me to sit under the table with blankets over it and
study by a candle after taps. Fortunately, I was never detected by
the tactical[19] officer at his inspection. At the end of the term, I stood
number eleven in the class. The first captain[20] asked me to take a
young lady, Miss Amie Erwin, to the graduating hop,[21] which was
the first that we were allowed to attend. While I saw little of her, she

was destined to be the wife years later of Colonel Robert R. McCormick,[22] who became one of my closest friends.

When the "makes"[23] were published after the graduation exercises in June 1889, I was fourth corporal. Nothing could have filled me with greater pride. We moved into camp, and the summer's work began. As yearlings, my class drilled the new cadets, and we felt our importance. My rank gave me the important assignment as a member of the color guard. I was transferred to "C" Company, where I was the senior corporal and, often, acting first sergeant. On one tour of guard duty, the other two corporals were caught trifling with sentinels and were reduced. My relief was off duty, and I was asleep, but I tried to perform my duties in a military manner, and, in any case, I would not have been involved with them. I attended a few dances, but I devoted practically all of my time to my military duties in camp. My tent mate, George H. McMaster,[24] was a superior character, and we became very good friends.

We returned to barracks on August 28 and resumed the academic course. The third class course was regarded as by far the most difficult, especially in mathematics. I lived in the fourth division on the first floor. The routine was now familiar, and I worked very hard. When the class of 1892 met to organize, I was entirely ignorant and unsuspecting of any plans. I was immediately nominated and elected chairman. When I called for nominations for class president, I was astonished to discover that two intense factions had carried on vigorous campaigns. One nominated Prince, and the other nominated Whitney.[25] The feeling was intense, but the Prince faction expected to win. When the vote showed the election of Whitney, the Prince faction were very resentful and never fully accepted Whitney. I was nominated and unanimously elected vice president with no opponent. The result has been that, after graduation, the class has looked to me for leadership. My class standing at the end of the term was number thirteen. Again, the climactic event of publishing the makes after graduation took place, and I was appointed senior first sergeant and assigned to "A" Company. This was one of the proudest moments of my life. My class at once left on furlough until August 28.

The balance due us on our accounts was paid to us. I had been elected chairman of the committee to arrange for the attendance of the class at the theater in New York and for the class dinner. The money for both was paid by the members of the class to me surreptitiously, although we received our balance, which I used for my expenses. We had selected the play by DeWolf Hopper[26] and the Murray Hill Hotel for the dinner. I knew nothing of such affairs, but both were a great success. At the dinner, the class prophet predicted a social career for me because I had refrained from social activities.

In order to save the money paid to me at West Point for use at home, I bought a steerage ticket to Jacksonville. After reaching Jacksonville, I then took a river steamer to Astor and went from there by train to Astatula. I spent the summer working in orange groves for small wages and improving the house in which my parents lived. At the same time, I helped my mother with the cooking, the washing, and the housework. A roster of the company was sent to me, and I memorized the roll. At that time, first sergeants called the roll rapidly from memory, alphabetically, and were compelled to catch the answer or failure to answer. The first sergeants competed with one another as to which company roll call would be completed first.

Again, I returned to New York City by steamer. On reaching West Point, on the *Mary Powell,* we put on uniforms and took our places for the first formation. I found no trouble in calling the roll and in detecting absentees. I occupied the first sergeant's room, which was the right-hand area room on the second floor of the Second Division. My roommate was George C. Barnhardt,[27] who was a very congenial friend. He was also the company clerk. At that time, guard mounting was immediately after breakfast. The company first sergeants marched their details to the formation and remained throughout the ceremony. This prevented me from studying before the first class at eight o'clock. Other duties of the first sergeant also occupied much of my time. The professor of philosophy had a theory that first sergeants could not study properly and perform their duties, and he acted accordingly. One first sergeant, Prince, was reduced because of poor class standing. I did lose grades, and at the end of the term, my standing was twenty-three. However, the pride and experience of being first sergeant of "A" Company were a rich recompense. I had some trouble with first classmen who, according to custom, transferred to "A" Company and tried to do as they pleased. I reported them for being improperly dressed at reveille and expected to have to fight. However, they realized that I was acting from a sense of duty and soon conducted themselves properly.

Although I attended a few hops, I felt that I had little time for anything but my military duties and my studies. It was during this year that Dennis Michie and Prince, with a few others of my class, took the lead in organizing a football team and playing the navy on the cadet parade ground. West Point lost badly that year[28] but defeated the navy at Annapolis the following year. Football, class yells, and spirit marked a change in discipline and in the relations between cadets and officers. During the first two years, practical military training consisted of infantry, light artillery, and coast artillery drill. In the second class year, we began riding in the riding hall, which included bareback riding; dismounting and mounting at the trot and gallop; and mounted drill. This was especially enjoyable, though I was not good at the acrobatic work. Gymnasium instruction continued, but I was not agile in the machine exercises. We had fencing, which I liked, especially the broad sword. I also liked boxing and did fairly well.

When the new makes were published after graduation exercises in June 1891, I was appointed first captain and, of course, assigned to "A" Company. I think this was the proudest moment of my life. The text about David came back to me, and I thanked the Lord for his blessing and being with me. The routine of camp and barracks life was very different as first captain from the previous years. The academic system, however, remained the same. When the section was seated for a recitation, the instructor had each cadet draw an enunciation, which he placed on the blackboard, then standing at attention until called on to recite. One day, in engineering, I drew an enunciation on how to run a traverse in surveying. I knew it well and filled the board with the process. Almost as soon as I began to recite, the instructor interrupted and asked me a question about some detail far along in the process. I replied: "We'll come to that presently, Lieutenant." He said: "We'll come to it right now, Mr. Summerall." Of course, the section was much amused at my expense. He saw that I knew the subject but wanted to demonstrate that I did know it by telling the class. I especially enjoyed the calculating of the strength of members and drawing the plans for a highway bridge. I enjoyed all the work of the first class year, even the higher mathematics in the course.

I took my duties very seriously, and in discharging them, it sometimes became necessary to report classmates, much to their resentment. A few men challenged me to fight, but before we could meet, they apologized. I did not believe that a cadet officer should have to defend the discharge of his duties by personal combat, but as such was the custom, I had no choice. I was not a fist fighter and would probably have been badly beaten by some, but that was a chance I had to take. I was not happy in my office, but the responsibility developed character and moral courage. I believe that the experience laid the foundation for my future career.

Cadets seldom raised their class standing after the second year, but I was able to graduate number twenty-one in a class of sixty-three. Cadets who had no demerits were given two days furlough at Christmas. I had taken the daughter of a wealthy family in New York to some hops, and they invited two classmates and me to their home. This was my first experience with wealth. They took us to the theater and gave us elaborate dinners and lunches.

During this year, football teams from different colleges played our team at West Point. Their presence in the mess hall, their college yells at meals, and the nonmilitary atmosphere introduced by them were in strange contrast to the ancient customs of the academy. When our team returned from defeating navy, it was met at the train station by the Corps of Cadets and borne on their shoulders with such yells and pandemonium as the hills of the Hudson had never seen. We realized with different emotions that, whether we liked it or not, a new era had been ushered in. Many old ideas died, and many new ones were born. The effect on some of the athletes was marked. One of my closest friends during the first two years became an undisciplined bully. I could have nothing to do with him, and I spoke to him only officially for nearly two years.

This period from 1888 to 1892 saw other marked changes at West Point. The full dress coat was discarded for wear at all times outside of the room, and the blouse was substituted except for ceremonies and chapel. The turnover collar on the full dress coat was replaced by a standing collar inside the coat collar. The old Civil War–era Springfield single-loading rifle was replaced by a magazine rifle. A breech-loading rifled gun with separate powder charge and projectiles replaced the muzzle-loading field gun of the artillery. Infantry drill was much changed, and other supposed improvements followed. The old academic building was razed and a new and larger stone structure erected. A gymnasium was constructed where the mess hall now stands. The size of the corps remained, however, at about 240 or 250.

During the spring of 1892, the tailors and equipment dealers came to outfit us. Then, the great moment of graduation came, and we were awarded the coveted diplomas. We lost no time in changing to our civilian clothes and leaving. Our joy far exceeded our regrets. My wealthy friends invited four of my class and the lower class to their beautiful home at Tuxedo Park, where many young ladies made up a house party. After a week of lavish entertainment, I intended to sail from New York for Jacksonville on my graduation leave. But the lady who was the real head of the family of my wealthy friends insisted on providing me with a ticket by rail at much greater expense. She had come to look on me as a sort of grandson, and I owe her many kindnesses.

Chapter 4

"We bid farewell to cadet gray and don the army blue"

A new life now dawned not only for me but for my impoverished family. It did not occur to us that my salary of $116.67 per month as a second lieutenant was not to be used for the benefit of all. This, to us, was great riches. We had never dreamed of anything like it, and it meant what we had never known—security. I knew nothing of what my living expenses in the army would be, but I determined that the greater part of my salary would be sent to my parents. On graduating, I made my first choice for assignment, the First Infantry, stationed in San Francisco Harbor, California. When I received my assignment to the regiment, I wrote to the captain of "C" Company,[1] to which I was ordered, expressing my pleasure at the prospect of being with him and my hope that I could be of some service. He replied from the Cosmos Club in San Francisco, where he was on recruiting duty. He stated that he wanted to give me two pieces of advice and that, if I would follow them, I would rise to the head of my profession. They were never to play cards for money and to read military history. I may say that I did follow his advice and I did rise to the head of my profession. There may or may not have been some connection. I found that he was noted for not following his own advice.

My wealthy friend in New York whose granddaughter I had taken to cadet hops wrote me that she and her granddaughter would be at the Palace Hotel in San Francisco the end of September when I arrived to join my regiment on September 30. I left Astatula in time to spend a few days with them. The railroads gave reduced rates to army officers, and I was able to buy a day coach ticket to San Francisco for about $40.00. It did not occur to me to ride in a Pullman and spend the extra money. In order to see something of the army en route, I stopped a day at Atlanta, Georgia, to see the Fourth Artillery at Fort McPherson; a day at Omaha, Nebraska, to see the

Second Infantry at Fort Omaha; and a day at Salt Lake City, Utah, to see the Sixteenth Infantry at Fort Douglas. I was treated with great courtesy and invited to lunch at each place. At Fort Omaha, Lieutenant W. M. Wright[2] was adjutant. He and his lovely wife gave me my first glimpse of an army home and of army hospitality. They drove me to the station in their phaeton[3] drawn by two horses. He became a lifelong friend. In World War I, he was a major general commanding a division in my army corps. No one could have been more loyal to me.

I ate at the meal stations en route. The railroad cars were very uncomfortable, and I made frequent changes, but I was saving money much needed at home. On reaching San Francisco, I took a carriage to the inner court of the old Palace Hotel. As I alighted, I saw my friends waving to me from an upper balcony. I was their guest, and they had reserved a lovely room for me. This was a new world, and the glamour quite overwhelmed me. Immediately after breakfast the next morning, the card of Captain Thomas H. Barry[4] was brought to me. He was the first officer to call on me, and I felt it an unexpected honor. He then commanded a company in the First Infantry at Angel Island. In after years, he proved to be a loyal and helpful friend. He was one of the ablest officers in the army, and years later, when I was a captain on duty at West Point, he was a major general and superintendent of the U.S. Military Academy. During the day, other officers called, and I was invited to visit Angel Island and meet the officers, which I did. My assignment was to "C" Company, stationed at Benicia Barracks, thirty miles east of San Francisco. I called on the captain of the company[5] at the Cosmos Club. He was cordial and told me to use his furniture stored at Benicia Barracks. I learned that he spent most of his time gambling and was one of the best card players in the city.

After spending a few days in San Francisco and having my photograph taken in my officer's uniform at their expense, my friends took me to the Del Monte Hotel at Monterey for a few days. Because I had never seen such luxury and expense, I wondered how anyone could be so rich and generous. It was all the antithesis of anything that I had known. On our return to San Francisco, my friends left, and I reported for duty at Benicia Barracks. The company commander immediately told me that it was a horrible, lonely place and appeared in every way to try to discourage me. I was filled with everything and found the life fascinating. The duties consisted of an hour's drill each day and being officer of the day occasionally. We spent the evenings at the homes of the married officers, playing cards, but not for money, or singing with the young daughters of the commanding officer. Occasionally, we went duck and goose hunting several miles away. A classmate of mine, a captain whose family would not live there, and I formed a mess with a Japanese cook. Our expenses were less than $30 per month apiece. I was thus enabled to send home more money than the family had ever seen.

Almost immediately after reporting to Benicia Barracks, I received invitations to teas for debutantes and dinners in San Francisco. I had bought a dress suit and a cutaway suit with my outfit at West Point. The fare to San Francisco was small. I went with my classmate to a large tea at a wealthy home. A young lady who was assisting took charge of me after I greeted the hostess and introduced me through several rooms. After she excused herself, I decided to leave. On my way through a room, I noticed a couple under a palm tree in a corner, talking earnestly. I knew the young man at once to be one of the boys with me at Dr. Porter's. I asked him if his name was not Brooks Jones. He replied resentfully that it was. I then said: "I am Charlie Summerall." He greeted me most warmly and introduced me to the young lady and to many of his friends. He explained that he lived with his bachelor uncle, Winfield S. Jones, and took me home with him.

His uncle welcomed me and became the best friend I ever had. He came of a distinguished family with prominent relations among the leading Virginia families. Mr. Jones was the vice president and manager of the Security Savings Bank. While not asserting his status, he was the social arbiter of San Francisco. He had never married, had a good home on Hyde Street, and lived well. He asked me to make his house my home and to stay there whenever I went to San Francisco. This opened the way to another social world, where I met the nicest people in the city, who invited me and my fellow officers to their homes and entertainments. A series of dances by the Cotillion Club was the chief social event. The cost of membership was high, but my friend in New York insisted on paying for me. Mr. Jones was very musical, and during the next four years he took me to many concerts and operas as well as having musical parties at his home. Thus, I learned much of value in the musical world.

At Benicia, I had much time for reading, and I obtained a number of books on military history, especially of the infantry, and the art of war. I wrote a paper for the officers' lyceum[6] on the new Krag-Jorgensen magazine rifle. I conducted the post school for enlisted men and thus came to understand them and their needs. Some of the officers had taken part in the Wounded Knee[7] Indian campaign in 1891. As there was no prospect of another Indian war, they pitied us for missing our last chance for fighting.

There was much drinking in the army and in the social circles I frequented. I had never taken liquor and did not drink wines and champagne at the dinners I attended or at the cotillion suppers. On Columbus Day, the Roman Catholic church in Benicia had a military mass. A platoon of soldiers from Benicia Barracks and the officers occupied the chancel. After the mass, a large lunch was served in the refectory of the monastery. Much wine was served, and I drank with the fathers. I soon felt numb in my legs and was sure that I could not stand when the lunch was over. I felt panicky at getting drunk at the priests' table and suffered great stress of mortification. My mind was clear, and I could talk naturally. However, when it was over, I rose and walked easily. It was a lesson, and I did not drink any more. Throughout my life, I have made it a rule not to drink any kind of alcohol or use tobacco in any form. People thought just as much of me.

It had been very difficult at West Point for me to decide in which branch I preferred to serve on graduation. All had attractions, but my inclination was for the artillery. Cadets made selections according to class standing, with the priority order of engineers, artillery, cavalry, and infantry. Because there were few vacancies in the artillery, those of my class who graduated high enough to win it were mostly assigned as additional second lieutenants. The cavalry enjoyed some prestige on account of the horses and increased pay as well as the Indian campaigns, especially the Custer massacre. Initially I considered the cavalry but later requested the First, Sixteenth, and Fifteenth Infantry because the First was in San Francisco, where I wanted to go. Promotion at that time was by branch. The cavalry and the infantry had more rapid promotion because of the Indian service. Promotion in the artillery was very slow.

On joining my regiment, I soon found a bitter prejudice among the branches. Officers serving in the infantry and the cavalry did not like each other, and both detested artillery officers, who were called *coffee coolers* because they were stationed in the harbors of cities. The artillery called itself the *scientific arm* and looked down on the infantry and cavalry. My contacts with the artillery at the Presidio of San Francisco, largely through the daughters of the commanding officer, General William Montrose Graham,[8] made that service very appealing. General Graham, a brigadier general by brevet[9] in the Civil War, was a colonel, but he was always called general, and he signed himself as brevet brigadier general. He was probably the most feared officer in the army. With a brilliant record as a light battery commander in the Civil War, he became what was improperly called a *martinet*. He was uncompromising in high standards of duty, discipline, and training, and he probably had the most efficient regiment in the army. Because there was no Class B system[10] then, many officers from the Civil War lingered when they were utterly unfit to serve. As far as possible, they obtained detached service. They would not retire voluntarily. These men found service under General Graham very unhappy and disliked him accordingly. Capable and industrious officers had no trouble with him.

I had the temerity to call at General Graham's office and tell him that I wanted to transfer to his regiment, the Fifth Artillery. He was cordial and told me that when my application was referred to him, he would ask a friend of his in the War Department to have it approved for a vacancy then existing. My application was a great shock to the officers at Benicia Barracks. A few minutes after it was filed with the adjutant, a captain came to see me in great distress and exclaimed: "My God, man, if you have the instinct of a soldier, don't transfer to the artillery." A number of infantry and cavalry officers said that I was committing professional suicide. The First Infantry, of which I was very fond, never forgave me.

The attitude of the older artillery officers toward the young lieutenants was formal instead of familiar, as with the infantry. As soon as I appeared at the Presidio, some of the old battery commanders told me to ask General Graham to assign me to their batteries. When I reported to the general, I told him this. He smiled knowingly and said: "I shall assign you to Lieutenant Reilly's battery." Here, without my suspecting it, was another evidence of the destiny that shaped my career. I had never heard of Lieutenant Reilly,[11] nor had he heard of me, but our lives were linked from that moment till the climactic end when he was instantly killed on the walls of Peking, observing my guns of his battery blowing open the great gates of the Imperial City. He had been an enlisted man in the Civil War and a first lieutenant since before I was born. I at once saw in him a remarkable personality. He was able, enthusiastic, and dominant, commanding the best battery on the post. When asked how he always had such good noncommissioned officers, he replied: "I make them." He was entirely wrapped up in his profession and his lovely wife and children. Through his wife's family, he had taken a year's leave to be an executive in the Pullman Company with a view of resigning from the service. When he returned, he said that no amount of money would induce him to change the army for a civil career.

Reilly commanded Battery K, Fifth Artillery, which manned an eight-inch muzzle-loading rifled battery, covering with its fire the inner harbor and a battery of fifteen-inch smooth-bore, muzzle-loading guns, commanding the Golden Gate. I had learned the service of these pieces at West Point and had been trained in firing them. Because I was the only other officer in the battery, he set me to work plotting the last season's target practice and determining the centers of impact. This was readily done from the record of splashes and target at the instant of impact. He then had me determine the velocity of powder blended from several lots. This I did with the Boulanger chronograph,[12] which I had learned to use at West Point. I also oriented and graduated the azimuth circle of one of the batteries so that the zero indicated the true north of Polaris. I dismounted some large guns from the top parapet of Fort Winfield Scott and, later, mounted a battery of fifteen-inch smooth-bore guns at Lime Point across the Golden Gate from Winfield Scott. In the meantime, I learned practical mechanical maneuvers, cordage, first aid, and much else from the noncommissioned officers. Guard mountings, inspections, parades, and other ceremonies were in full dress, and the standard of military routine compared favorably to West Point.

I continued my reading of classics and military works and wrote papers for the officers' lyceum. There were many boards of which I was recorder, and I was often a member or judge advocate of courts-martial. All proceedings were written in longhand, as the typewriter was as yet unknown. Among my assignments was the repair of the water main that ran below high tide along the five-hundred-foot bluff on the ocean. Although my men and I worked day and night, the caving bluff carried away the line. We then abandoned the project. About this time, a modern twelve-inch battery was constructed, and my battery was employed to test it. I occupied the observation post near the water. While I was waiting for the fog to lift, a telephone call asked the condition of the fog, and I reported. A major who was in charge asked me the meaning of the call. When I told him, he said: "If anyone else wants to know, you report to me, and I will tell you the condition of the fog." This illustrated the attitude toward younger officers of the old men.

A year after graduation, I received my first "fogy,"[13] as it was the law then to increase the pay five years from becoming a cadet. My salary was now $133.38 per month. On going to the Presidio, I found that the club mess was too expensive for me. Several of the young officers had established a mess, called the *Scrub Mess.* These were the finest officers on the post, and I had known them as cadets. They allowed me to join them. The expenses were kept to $30 per month for the food and the care of our rooms and laundry. I was thus able to support not only my parents and my sister's family but also my brother and his family while he was studying medicine at Tulane University. The only clothing that I bought was a cheap blouse and breeches made by the battery tailor. After the winter of 1893–1894, I could not afford the fees for the clothing I needed for the cotillions, and I virtually dropped out of all social life in the city.

In 1894, I was detailed to duty with one of the light batteries[14] at the post. This gave me an increase in pay of $10 per month. I loved horses, and my service became more interesting. We made marches each summer and camped at Monterey for maneuvers and target practice on the Jacks estate, which covered many thousands of acres. Again, I did not dream that one day I would purchase this ground for the national guard artillery and that it would become a great training center. As engineer officer on the march, I mapped hundreds of miles of road and became familiar with the country and towns from the Coast Range to the ocean.

The lack of promotion was the chief topic of conversation among the older officers, and many had no hope. It was disturbing to the younger officers. According to tables that calculated promotion possibilities, I would become a major a short time before I retired at the age of sixty-four. This led me to consider transferring to the Ordnance Department, where promotion to first lieutenant would take place at once and in twelve years I would become a captain. Selections were made by competitive examinations in New York. In the early summer of 1894, I applied and was ordered to take the examination. While it did not appear to be very difficult, I was unsuccessful. Here again, destiny veiled my fate, for I would have been submerged in my career as an ordnance officer. I made two more efforts, but other competitors were selected.

I traveled in a day coach each way and thus saved most of my mileage money[15] for the family. On returning to San Francisco, I read at Cheyenne, Wyoming, that the American Railway Union intended to strike. My train was the last to reach San Francisco. Traffic was stopped on all roads west of the Mississippi. The next night, General Graham sent for me and told me that the Presidio garrison, the marines at Mare Island, and the infantry at Benicia Barracks would leave next day for Sacramento. He ordered me to secure steamers to transport the troopers and their supplies up the Sacramento River. We were to restore traffic on the Southern Pacific Railroad from Truckee, Nevada, to San Francisco. It would be a long story to tell how that night I secured ferryboats, loaded supplies for troops, and had the troops at the Presidio embark with horses for the cavalry, stop at Mare Island for the marines and at Benicia for the infantry, and reach Sacramento early the following morning. The railroad yards were filled with stalled trains, spoiling food, and stranded passengers. At once, a mail train departed for San Francisco. When it reached the American River Bridge, the fishplates, the steel plates used to connect sections of rail together, had been removed and the piles sawed through. The locomotive overturned, killing the engineer, the fireman, and five soldiers who were a guard on the tender. The entire train was wrecked. In the meantime, the mob in the streets had thrown missiles at the soldiers, who then fired into it, killing one man. The state militia had proved entirely useless.

I had the bodies of our men properly cared for when the locomotive was raised. They were badly decomposed. The government allowed only $10 for a coffin for a soldier, but I could not buy one for less than $40. Our officers contributed the difference. They were buried in the cemetery at the Presidio, and General Graham had the inscription "Murdered by Strikers" placed on the stone over their graves. This caused a violent protest from organized labor, and the War Department ordered the stone removed. The criminals who wrecked the train and killed the soldiers were never punished. The strike was led by Samuel Gompers and Eugene Debs, and many years later I was to play a part in the dramatic funeral of the former. For six weeks the troops guarded trains and tracks and restored traffic.[16]

Shortly after I returned to the Presidio, General Graham appointed me regimental adjutant. While his action was legal, it was almost the invariable rule of the service to appoint senior first lieutenants to this important position. Immediately, the older officers advised me to decline it and said that, if I accepted, I might be ostracized by the regiment. I did not want the position, but I did not see how I could refuse to obey an order. The attitude of the older officers seemed to me to be wrong. When the War Department received the order appointing me, it dispatched a letter to General Graham advising him to revoke the order, which he did. Later, he appointed me post exchange officer, the principal business of whom was the sale of beer. I came into intimate contact with the drunkenness of the men, promoted by me since, as the post exchange officer, I permitted the sale of beer. These duties, however, made money for the company fund. The steward had been in the exchange for years, and it had always been customary to let him deposit the money collected at payday. Many officers gave him their paychecks to be cashed at the bank. He had always been honest and was trusted implicitly. It would have given him great offense for me to relieve him of this duty when he handled the cash receipts and turned them in to me daily. On one payday, he had not returned from depositing the money when I went to check to see if he had the receipt. I at once notified the police. He was found drunk with all the money, except $300, in his pocket. It was necessary for me to refund this amount from personal funds, which I did by borrowing it from Mr. Winfield Jones. In due time, I was able to pay him. This taught me an invaluable lesson, and I have never since trusted money to an unbonded person.

In August 1896, after returning to the Presidio from a practice march and target practice, I received a letter from my mother recounting my father's death from malarial fever. This was the first bereavement in our immediate family, and it affected me deeply. The expense and the distance made my going home unthinkable. He was buried in the lot that I had purchased in the cemetery in Eustis. A few years later, when I was able, I had a suitable stone placed over his grave. My father was always kind, loving, and helpful. He was very proud of me. He had sacrificed and done all he could for us. Without an opportunity for an education, he had broad knowledge and many skills. He was well versed in medicine and law. His life on the plantation made him familiar with farming and livestock. He could design and build a house, and he knew the art of painting, by which he sometimes made a living. He could build and repair carriages and wagons, and he understood blacksmithing and cabinetmaking. He had the highest integrity and understood the mercantile business. He wrote and talked well, and his reasoning was sound. I can never explain his lack of success unless it was due to complete unselfishness. He and my mother were wholly devoted to each other. He could not bear it to be away from her, and he could not long have survived her, just as she did not long survive him.

In a very short time after my father's death, destiny began to develop the most momentous part of my life. Lieutenant John D. Miley[17] and his wife were a young couple who occupied an apartment in the "Corral," where a number of young officers, including myself, had rooms. Miley was the most superior officer on the post, and his wife was brilliant and entertaining. They had two young children, and the young officers often called on them in the evening. We were told that Mrs. Miley's sister, Miss Laura Mordecai,[18] whose father was Colonel Mordecai[19] of the Ordnance Department, would arrive soon for a visit. People on the post who knew Miss Mordecai all spoke of her admiringly. The morning after she arrived, I met her with the Miley children at guard mounting. Knowing who she was, I introduced myself, and she spoke to me most graciously. I was deeply impressed by her sweetness, loveliness, and daintiness. I thought, How small she is. In later years, I learned that her mother and others had commented on her being so small. After a brief conversation, I went on, but the incident clung to me. I called on her at once and met her occasionally. On account of my father's death, I did not go to the post hops, but the officers whom I saw said she danced beautifully. One evening, I worked late on the proceedings of a court-martial, and when I finished, I went out for a walk. On passing the Miley apartment, I saw her sitting on the steps, and I asked her to walk with me. Without intending to do so, we walked along the beach to Fort Winfield Scott. The time and the distance seemed short to me, and I have no doubt that she enjoyed being with me. She played the organ for the post chapel services, where I sang. I soon found that she was deeply religious and was interested in welfare work.

Soon afterward, the regiment was transferred to New York Harbor, and my battery was ordered to Fort Hamilton. Lieutenant Miley remained at the Presidio to continue the mounting of modern guns in which he had been engaged. In saying good-bye to Miss Mordecai, I felt that she had entered my life for always, though nothing was said or done to indicate it. We corresponded at once. Later, I visited my classmate Lieutenant Tracy O. Dickson[20] at Springfield Armory, commanded by Colonel Mordecai. She visited friends in Bay Ridge, Brooklyn, and I called on her there. I realized that she meant more to me than anyone else had ever done, but my financial responsibilities precluded any thought of marriage.

The transfer of the battery to Fort Hamilton included only the officers and men as we took over the horses and materiel left by the battery of the First Artillery, which exchanged stations with us. We traveled by special train, which took seven days to make the trip, early in October 1896. I took a three month's leave in December and spent it with my family in the house in Eustis. It was the last time that the family were all together, and, thanks to my pay, we lived very well. It was then that my brother made plans to try to establish a medical practice in Atlanta, Georgia. Of course, I furnished the money for everything until he could earn a living. This did not happen. I occupied a set of quarters at Fort Hamilton in which I could afford to heat only one room. I took meals with Lieutenant and Mrs. Gatley,[21] who lived next door, at a small cost. I made no attempt to revive my acquaintance in New York and seldom left the post.

During 1897, the "Yellow journals," mostly the Hearst papers, began to stir up feeling for a war with Spain over Cuba. We could not take it seriously. During the summer, the battery marched to Tyringham Valley, Massachusetts, for target practice. I was quartermaster and found some difficulty in securing supplies and campgrounds. At one place, I was compelled to buy dried clover instead of hay. The captain was displeased and said the horses would not eat it. Of course, they ate every particle, and it was far better than hay. At another place, we camped in a Quaker community. The men did not like our spending Sunday with them, but they gladly accepted pay for the campground and supplies. The women and girls visited the camp and appeared to enjoy the sight of the men. They were dressed in black with no jewelry. One camp was at South Egermont, near Great Barrington. A family with whom several young ladies were summering as guests invited the officers to dinner. The young ladies asked me what I thought of the Northern people. I told them that they were just like the people in Virginia and South Carolina. A class of people is the same everywhere. It rained so hard the next morning that we could not cook breakfast. I rode ahead to Great Barrington and asked the proprietor of the Red Lion Inn if he would serve breakfast to our 125 men and officers, intending to pay him from the battery fund. When he said yes, the battery arrived, a nice breakfast was served, and he would not accept payment. Our target practice was conducted in a Shaker settlement over some fields and wasteland. It was a pleasant outing for all in spite of a great deal of rain. We camped in several small communities that have now become dense cities, and a horse battery could not now march through New York and along Broadway to Yonkers.

The tension with Spain mounted. In the early part of 1898, the Spanish cruiser *Viscaya* visited New York and anchored near Fort Hamilton. I was officer of the day and fired the return salute from the old fort. She made a short stay, much to the relief of everyone, as it was feared that she might be damaged by lawless persons.

Chapter 5

Remember the *Maine*

The post had arranged for a large dance on the night of February 15, 1898. I was on the hop committee. We had invited the navy officers and ladies from the Brooklyn Navy Yard, and they had accepted. Then the country was electrified by the news that the battleship *Maine* had been blown up in Havana Harbor.[1] We discussed whether to cancel the hop or to ask the navy whether they would come, but we decided to do neither. All of the navy came to what was a large dance. The next day, reporters from the New York papers came to find out about the dance. When told by the butler at Colonel Ramsay's quarters that Colonel Ramsay[2] was out, the reporter asked who was second in command. The butler replied: "There ain't no second in command at Colonel Ramsay's post." We had known no details about the *Maine* and would have canceled the dance had we known of the casualties and the loss of the ship.

When war was declared, General Graham was a brigadier general commanding the Department of the Gulf with headquarters at Fort Sam Houston, San Antonio, Texas. He was made a major general and at once invited me to become his aide-de-camp. I was detailed April 8, 1898. I was very busy with him at San Antonio and in Atlanta, Georgia, putting the defenses of the Gulf and the South Atlantic in some state of efficiency.

On reaching Jacksonville with him on an inspection trip on April 21, 1898, I received a telegram saying that my mother had died the previous night. I went home at once in time for her funeral. It would be impossible to express what my mother had meant to me and the sorrow that her death brought. Whatever I have done worthwhile in my life has been due to her and her influence. Her loving, tender care in childhood; her understanding and encouragement in maturity; her inspiration of all that I ever did; her ambition and hope for us; her selfless sacrifice that we might live in the darkest days of poverty; her ambition for us to succeed; her intervention whenever we found difficulties; her strength as a refuge in our troubles; her fine intellect; her noble character and brave spirit made our lives and our success. Her beautiful letters came frequently and refreshed me unfailingly. She taught me all that I knew before going to Dr. Porter's, and I found myself thoroughly grounded. Her appeal to Dr. Porter to take my brother and myself was the turning point in our lives. She was never despondent and never complained. In the most abject poverty, she wrote the most beautiful verses. To have been blessed with such a mother and such a wife as mine is more than anyone could deserve. She asked only that we should place on the stone at her grave the words: "She hath done what she could." This was done. She was laid by my father and her dear sister, Susan, and their gravestones are alike except for the inscriptions.

General Graham was assigned to the command of the Second Army Corps at Camp Alger, Virginia,[3] and took command on May 19, 1898. General Graham found one regiment in camp with men dying of typhoid fever. There were no hospital facilities, and the sick men lay on straw on the ground in their tents. At that time, no one knew the cause of typhoid or how to prevent or treat it. The state militia poured in in a pitiable condition. The men were poorly clothed and fed, and the officers had no training for their duties. The federalized and instructed national guard was unknown then. The task of organizing, training, and caring for these thousands of helpless and sick boys, most of whom were under age, was over-whelming. The provision of water presented almost insuperable difficulties. That some order and sanitation with reasonable food and shelter were obtained was due in the greatest measure to the indomitable spirit, high standards, and broad experience of General Graham. His staff labored incessantly. Three of them had typhoid, and one died. In the end, the general also had typhoid. It is remark-able that any of us escaped.

An incident of unusual importance occurred when the Third Virginia Regiment created a considerable disorder one night. Major General M. C. Butler[4] of South Carolina, in command of the divi-sion to which the Virginia regiment belonged, rode with his staff to quell the disorder. The men jeered him and refused to obey. He confined the regiment to its camp and placed a Connecticut regi-ment to guard it. The Virginians bitterly resented this treatment, and the colonel appealed to the corps commander for a court of in-quiry. I was made recorder of the court, which meant that I was to conduct the case and present and record the evidence. The court consisted of three colonels. When the court met, I noted a number of distinguished-looking men in the room. The colonel of the regiment introduced them as his counsel. They included the senators and congressmen from Virginia, the attorney general of Virginia, and a very able attorney from Washington, who conducted the defense. As a young second lieutenant, I was at a great disadvantage in the technicalities that might arise. Immediately, General Gobin,[5] a prominent lawyer and the adjutant general of Pennsylvania, who commanded a brigade in another division, asked me in a whisper if I would like him as assistant to the recorder. While such a thing was without precedent, I introduced him accordingly. I had employed several men who were court stenographers in Washington to record the proceedings and the evidence. The record finally filled several volumes. Many witnesses were called by General Butler and by the regiment. I endeavored to conform to the procedure of a court. General Butler often became so violent in his attack on the regiment that I would ask the court to recess to let him become quiet. During this time, when the Second Army Corps was transferred to Camp Meade, Pennsylvania, I took the findings of the court to the corps commander there. The situation was very embarrassing. It would have had a bad effect to publish a censure of the Virginia regiment or to fail to uphold the authority of General Butler. But so much time had passed since the incident and it had been so thoroughly aired that the purpose of the inquiry had been accomplished. No final action was ever taken on the proceedings.[6]

While we were at Camp Alger, the expedition sailed to Cuba and fought the battles of San Juan Hill and Santiago. Colonel Miley, whom I had known so well at the Presidio, was aide to General Shafter,[7] who was colonel of the First Infantry when I was in it. All accounts credited Colonel Miley with whatever was good in the campaign. A number of the troops at Camp Alger were sent to Puerto Rico. We had hoped that the Second Army Corps would be sent, but the war was short, and the corps was used to organize and train troops.

The problems at Camp Meade were much the same as at Camp Alger, but much progress had been made in solving them. Hospital tents were provided, and for the first time some women nurses were employed. Because the troops felt that the war was over and wanted to go home, many men went absent without leave. Just before leaving Camp Alger, one entire regiment left en masse and never came back. The politicians made our problems more difficult. President William McKinley reviewed the corps at Camp Meade. During the review, he expressed surprise at seeing General Graham there and said that he had ordered that General Graham be sent to command in the Philippines when troops were sent there. It was evident that the adjutant general, who controlled the army, had paid no attention to the president's order. General Graham had reached the retirement age early in the war but was continued on active duty because of the need for his ability. He was retired and relieved of the command of the corps on November 11, 1898, and I was relieved as aide-de-camp. At this time, he suffered a severe attack of typhoid fever.

General A. C. M. Pennington,[8] commanding the Department of the Gulf, with headquarters at Atlanta, invited me to become his aide. I had been there with General Graham when the war began, so it was most agreeable to see again my Atlanta friends. I was fortunate in contacting a schoolmate at Dr. Porter's, Elliot Jennings. We shared a room at a new apartment building. Some nice people invited me out, but I went very little.

Chapter 6

The Little Brown Brother

Early in 1899, the Philippine insurrection became serious, and Reilly's Battery was ordered from Fort Hamilton to Manila.[1] Captain H. J. Reilly, who was in command, requested my assignment to the battery. I was relieved as aide to General Pennington on March 2, 1899, and joined the battery en route at Ogden, Utah. Thus, a new phase of life was ushered in by destiny. I had just been examined for promotion to first lieutenant at Fort Monroe when my orders came. The promotion was effective March 2, 1899, but I did not receive my commission until later. This increased my pay to $166.67 per month, which greatly helped me meet my financial obligations, including assisting my sister's family.

During all these hectic days of the war, I corresponded intensely with Miss Mordecai. She visited some relatives in Washington while I was at Camp Alger, and I saw her a few times. There could be no thought about our future, for I was not in a financial position to marry. She accepted the conditions bravely. Our letters continued to be frequent after I left for the Philippines. She wrote the most beautiful, inspiring letters that I had ever seen.

The other lieutenants in the battery, selected by Captain Reilly, were L. R. Burgess[2] of my class at West Point and Manus McCloskey[3] of the class of 1898, who had served with the battery in Cuba. They were most capable officers, and we were all warm friends. The battery was traveling in two special trains. One carried the personnel and baggage and the other the horses, mules, guns, caissons, and wagons. I was at once assigned to take charge of the latter to San Francisco.

The movement of troops to the Philippines marked a new era for our country. There was drama in the voyage. Men sang the songs of war and of farewell. None realized how many would never return. All were buoyant and yet filled with sadness at going so far from loved ones. After some delay at the Presidio, the animals and guns were loaded on a freighter, and on April 8, 1899, the personnel embarked on the *Newport,* a small coasting steamer that was entirely unsuited to the voyage. She was crowded with Reilly's Battery of the Fifth Artillery, Taylor's Battery[4] of the Fourth Artillery, and a battalion of marines. En route, the two batteries on board were organized into a battalion under Major Tiernon[5] of which I became the adjutant. After much delay, we reached Honolulu in ten days. Following repairs, we proceeded, reaching Manila twenty days later in a typhoon. I was very sick much of the time.

Before we left the states, much recrimination had appeared in the newspapers about the campaign in Cuba and the British campaign in South Africa at that time, which had been called the *graveyard of reputations.*[6] In discussing these circumstances one day on the ship, Captain Reilly suddenly said to us: "Gentlemen, there must never be anything to explain in this battery." His steel blue eyes were penetrating, and his words never ceased to ring in our ears as an unexpressed motto when each of us constituted a separate command.

We sailed past Corregidor at night in a terrible gale and anchored several miles from Manila because there were no docks. The next morning, I went with Major Tiernon to report to General Otis,[7] whose headquarters was in an old Spanish palace. I at once met Lieutenant Fred Sladen,[8] who was General Otis's aide and had been two years ahead of me at West Point. He said: "Well, I suppose you are all glad to be here." I replied that we were. He answered: "We all want to go home." I was to realize the meaning of his words in the homesickness, the diseases, and the loneliness that soon made our men, and the volunteers who had gone first, desperate to come home. The men were brought ashore in cascoes, or large native barges, and went into camp on the shore of the bay. When the ship with the animals and guns arrived, we were sent by platoons of two guns each to occupy separate parts of the south line, extending from Tagig on the Pasig River to the bay.

I went to Pateros, where I found a company of the Twelfth Infantry under an officer who was a cadet with me at West Point.[9] The rain continually poured in torrents, until the country was under water. Very little food or forage could be moved, so the animals were sent back to Manila. We had tents, but the men were soaking wet all the time. The rations consisted almost entirely of canned salmon and hardtack. Many, including myself, were at once attacked by dysentery, and others had malaria. A doctor walked the lines each day, but he could do little for us. The men tried to keep up their morale, but all were greatly depressed. We did not see the other two platoons, which were several miles away. The insurgents were not active, and there was no intention of stirring them up.

After some weeks, we were returned to Manila. My platoon was ordered to Calamba, via the Pasig River and the Laguna de Bay, about forty miles from Manila. We took the men, mules, and guns on cascoes towed by tugs. I reported at Calamba to Colonel Kline,[10] commanding the Twenty-first Infantry. He was much irritated that I was to relieve Lieutenant Alston Hamilton,[11] whose platoon had long been with him. Here again, I found that the infantry distrusted the artillery but liked an officer who gained its confidence. It was necessary for me to overcome the colonel's prejudice and that of the regiment. My command consisted of two 3.2-inch guns, served by my platoon of about thirty men, and two Hotchkiss revolving cannon,[12] and two 3-inch mountain guns, manned by details from the Twenty-first Infantry. The line extended in an oval in front of Calamba, with the flanks on a lake, the Laguna de Bay. The guns were posted in pairs at different points so as to command the entire front. Men and animals not on duty at the guns found shelter in the town of Calamba.

The insurgents were very active. While there was some firing during the day, at night they would approach the town and make rather strong attacks. My guns were very effective both day and night and were quickly moved to repel night attacks. I soon gained the confidence of Colonel Kline and his regiment. Lieutenant Colonel James Parker[13] of the cavalry was attached to the Twenty-first Infantry. At the Presidio, I had known him distantly as a captain of one of the cavalry troops and was very fond of his lovely wife and their attractive children, one of whom was to be an important part of my command many years later. Colonel Parker, though very reserved, visited me frequently to discuss conditions. He was very tall, had a striking personality and high courage, and was a real leader.

The command formed a plan to attack the insurgents across the San Juan and San Cristobal Rivers and drive them some distance so as to rid us of the frequent night attacks. It was well carried out, and my men served the guns with great accuracy and effect. Colonel Parker was especially pleased by the quick destruction of a sugar mill from which much fire came. When the action was over, an officer asked me if he could use my bamboo cart, used to haul ammunition, to carry the bodies of the men who had been killed back to Calamba. Because it was my first attack, I was so exhilarated that I did not think of anyone being killed. I was the more surprised because my men were entirely exposed, standing to serve the guns in the open while the infantry were lying down behind the rice paddy embankments. It made a great impression on me to see these bodies lifted and taken away in my cart. It was the first time I ever saw men killed.

When we joined the regiment, it was eaten up with malaria and dysentery. My men soon grew worse, with approximately one-third in the hospital at all times. The medical care was excellent, and our food improved by having fresh meat, bread, and vegetables delivered by boat from Manila. All were sheltered in native nipa and bamboo houses. The morale was high. The men wore beards and became hard and tough. While there was much to depress them, no one thought of a pass to go away.

One day, a young officer named Stockley[14] arrived on a boat with two pneumatic dynamite guns[15] to be tested against the insurgents. We placed them in front of an enemy trench and fired. The charge of dynamite made a terrific concussion, and the occupants of the trench ran away. All successive shots went wild in spite of all our efforts to hit the trench. The insurgents jeered at us and returned. The heat in the gun made the propelling air pressure so variable that accuracy was impossible, so Stockley took the gun back to Manila. We were soon to have tragic news of him.

After a few months, the Thirty-ninth Volunteer Infantry Regiment, commanded by Colonel R. L. Bullard,[16] destined to become one of the best friends and greatest influences in my life, relieved the Twenty-first Infantry. I had never known Colonel Bullard, and he had never heard of me. As soon as he assumed command, he issued an order relieving me of the command of the revolving cannon and the mountain guns, manned by the infantry. I immediately reported this to Colonel Kline, explaining that I had no desire to keep them but that I felt the safety of the position would be weakened by putting a strange officer and untrained men over them at once. I expected an attack on the new regiment that night. Colonel Kline protested to Colonel Bullard, and the order was revoked. No doubt, Colonel Bullard felt the prevailing prejudice against the artillery.

The Twenty-first Infantry marched out that day and sailed for Manila. As they passed along the street, I was shocked by the emaciated and weakened condition of the men. They were almost skeletons from disease and hardship, and many seemed to walk with difficulty. By contrast, the Thirty-ninth Volunteers, just from the states, looked strong and robust. That night, the insurgents welcomed the new regiment with an attack that reached the edge of the town. My guns were moved quickly to the point of danger and immediately covered the enemy with canister and shrapnel. They fell back shortly. This had a good effect on the new troops and especially on Colonel Bullard. The next day, he came to my shack and said that he proposed to drive the insurgents for a long distance along the Binan–Santa Rosa road, where they were in the greatest strength. When he asked my advice as to the plan of attack, we came to an agreement at once. The troops moved and occupied their positions before daylight. By necessity, my guns remained in the roads, as they could not be moved over the rice paddies. The infantry lay down behind the rice paddy banks and in the edge of the jungle.

Daylight broke suddenly, and the *insurrectos* promptly opened fire. My guns covered their trenches with shrapnel while the infantry delivered a heavy fire. I then blew down with shell some earthworks across the road, and we advanced. After his troops occupied the position, Colonel Bullard rode up to me and said: "I am surprised at you! I am surprised at you!" Thinking I had displeased him and he was going to censure me, I replied, perhaps somewhat resentfully: "What have I done." He replied: "You and your men stood in the road fully exposed while my men had cover, lying down." I said that we could not do otherwise and serve the guns and that artillery was always exposed. He was delighted with the performance of the guns and complimented me profusely. From that moment to his death, he was one of my best and most helpful friends.

Our advance continued throughout the day until we drove them, after frequent stands, beyond Santa Rosa and Binan, about fifteen miles. We bivouacked at Binan and had some skirmishing the next day. At one place, I indiscreetly rode beyond the outpost to investigate and had scarcely returned when a heavy attack drove in the outpost and brought the entire command into action. The return march to Calamba was made at night. My guns were with the rear guard as we expected harassing fire from the rear. The column at one place halted an unusually long time, and I rode forward to see what was causing the delay. As I passed a company, I heard a man say: "There goes that artillery officer. Now we will move." I felt that my reputation was made with the Thirty-ninth Volunteers. Colonel Bullard was kind enough to recommend me for two brevets for my performance of duty. There were no other awards at that time.

The insurgents were cleared beyond the San Juan and the San Cristobal Rivers, but they were in considerable force on the other half of our front and were well handled by General Malvar.[17] Colonel Bullard determined to drive them away. On January 1, 1900, he sent one battalion of infantry along the main road to Lipa, a second battalion to the right, and a third one to the left, to encircle a strong position at the Puente de Viga, which was an iron bridge over a deep, almost impassable ravine. My guns accompanied the center battalion. We had little difficulty until we reached the Puente de Viga, which we found strongly held on the far side. The fire of my guns soon dislodged the *insurrectos,* allowing the infantry to rush across. But we heard nothing of the flanking columns, which had found it impossible to advance through the jungle. On making camp, to our surprise, other troops appeared from the direction of Taal. An officer asked me if I had seen Stockley, who had brought the dynamite guns to Calamba. I replied: "No." It appeared that Stockley's father was financially interested in the manufacture of the guns and had his son commissioned and sent to the Philippines to test them. When they failed, Stockley was assigned to troops. Although he and one of two men had been sent ahead as scouts to Taal, nothing was heard of them. Years after the war, we learned that he and the men were killed and their bodies thrown into Lake Taal by the insurgents or the people of Taaland.

We soon fell in with Cheatham's[18] battalion, composed of men who remained from the Tennessee Volunteers when they were sent home. Colonel Bullard ordered Cheatham and me to attack a position in the direction of Lake Taal on the following morning. By enfilading the trenches with the guns at daylight, we forced the insurgents to retreat. Shortly afterward, we noted a column of troops coming from the direction of Taal. It was a regiment of volunteers under Colonel Anderson,[19] a cavalry officer. When Anderson rode up to Major Cheatham and me, he spoke to Cheatham but ignored me. Cheatham then introduced me as commanding the artillery, whereupon Colonel Anderson said in the most contemptuous way, without looking at me, "Artillery ——," using the filthiest word that he could utter. Again, I was the victim of the bitter branch hatred.

Anderson and his regiment moved on. We followed the next

morning and soon overtook Colonel Bullard with his regiment. I joined Bullard, who was quite annoyed at seeing Anderson, his senior, assume command of the advance against Lipa. We had not gone far when Colonel Anderson's adjutant rode back to us and gave me orders to take my guns and report to Anderson. When I replied that I was under Colonel Bullard's orders, Bullard told me to comply. Some distance ahead I found Colonel Anderson waiting for me. In the most friendly and cordial manner, he said: "I am glad to see you, Summerall. They are right ahead of us." His regiment was deployed and lying down under cover, but there was little firing and no evidence of attack. I located an enemy trench from which a good deal of fire was coming. There was also some sniping from trees. My guns soon had the range, and with shell and shrapnel we drove the enemy away. On reaching Lipa, I learned that a battalion of Bullard's regiment had made a dash to seize a large quantity of money that the insurgents were taking away. I followed with the guns and came on Bullard just after he had overtaken and captured the treasure. That night an officer said to me: "I hear Colonel Anderson is praising you too." He was very friendly to me the rest of his life.

The horses and men were exhausted. Horses fell in the harness during the long uphill climb to Lipa. The men were cadaverous from malaria and dysentery, but it was a rule in Reilly's Battery never to complain. I heard a young corporal say when the going was hard: "Fellows, I would not change places with anybody in the world." They were fierce fighters with unbreakable morale.

We pressed the march the next day and reached Batangas about noon. We had had little food and no forage since we left Calamba. I at once obtained both from a boat that was at the dock. On arriving, I ate a can of fois gras that had been sent to me at Christmas and that I carried in my saddlebag. At the same time, an officer borrowed my only piece of soap to bathe in the river. I never saw it again as I left before his return. The Fourth Cavalry had just preceded us at Batangas, and the colonel assumed command of all troops. In a few hours, I was ordered to accompany a battalion of the Thirty-ninth Infantry to a place called San Jose. The battalion commander and I, indiscreetly riding ahead of the advance guard, ran into a band of *insurrectos*. We leaped from our horses and began firing our pistols at them when they ran into the jungle. On leaving Batangas, the advance guard took the wrong road. The battalion commander recalled it while he and I took the correct trail and ran into the *insurrectos*. The battalion commander ordered the advance guard company to make a night march to Rosario while the rest of the battalion went into bivouac. The advance guard, harried by bolomen[20] all night, had a number of casualties. This so unnerved the young lieutenant in command that he committed suicide a few days later. Other troops came up the next morning, and the entire command advanced on Majayjay, which was reported strongly held.

The advance guard was composed of Macabebe[21] scouts, and I was ordered to accompany it. I found it commanded by Captain Lee Hall, a famous Texas ranger.[22] He asked what I wanted to do, and I said that I, with the guns, would follow and support the advance guard. He at once agreed and rode with me. We were ambushed several times, and there was a good deal of firing, but the guns soon drove the *insurrectos* back. There were some casualties among the scouts. Captain Hall and I became very friendly. He was a remarkable type of man with proven courage. His blond face was weather beaten and deeply lined. His eyes were like blue steel, and his hair and moustache were like blonde wires.

On reaching Majayjay, I was able to place the mountain guns in an advantageous position to fire on the defenses. The infantry deployed for the attack. When all was ready, we found that the *insurrectos* had withdrawn. The next day, the march continued to Santa Cruz on the Laguna de Bay. I accompanied Cheatham's battalion, which constituted the advance guard. His men helped me bridge deep ravines and move the guns over boggy places. As with the Thirty-ninth Infantry, the infantry would do anything for the artillery once the latter had won their confidence. To pull the guns through boggy ground, a whole company would man the prolonges[23] and pull as horses could not do. They built bamboo bridges and almost lifted the guns over ravines. We reached Santa Cruz at night, and I at once procured rations and forage from a boat at the dock. When the men and horses were cared for, I sought some rest in a nipa shack. I was wakened by odd sounds and found that the Tennesseans had moved in and were shooting craps by candlelight with gold pieces for stakes. I moved out. The next day, we embarked on cascoes with the men, horses, and guns and were towed to Calamba.

I had made my pallet in a nipa shack and was preparing to lie down when Captain Lee Hall came in and said: "Summerall, I have come to say good-bye." I asked him what he meant. He replied: "I am going to Manila on that boat tonight and resign and go home. I have been thinking a lot these last few days. Every day, I have been going along these trails with the insurgents shooting and killing some of my men, and some day it will be my turn. I am not going to let that day come. I will never go out there again." I tried to remonstrate by saying that I must continue to go and he could not leave me, but he was adamant. We shook hands with some mutual understanding, and he left. I realized that this was another case of men knowing by some intuition that the end of a life of danger and daring had come. Many years later, I read a biography of him that declared he left the Philippines because he was sick. I am probably the only one to whom he had confided. He was an unforgettable character. I can understand the story that, where he went, only one ranger was necessary to stop a riot. He was said to have killed a number of desperados.

This was our last campaign in the Philippines, for other work was soon found for Reilly's Battery. After some time, the three platoons were assembled at the Agricultural Grounds in Manila. The other two platoons had acted separately, supporting the troops operating in Cavite Province as my platoon had done in Laguna and Batangas Provinces. All had acquitted themselves creditably. Our losses had been few. Lieutenant Burgess had been badly wounded in the leg and sent to San Francisco to recuperate. Captain Reilly was well pleased with his officers and men. There was nothing to explain. I happened to hear Colonel Bullard telling him about me and saying: "He did not have to have a hammer to drive a nail." The battery at once began intensive training in gunnery, marching, and in the care of horses and men.

Chapter 7

The Land of the Dragon

In 1900, the uprising that broke out in China to expel all foreign devils startled the world.[1] International naval contingents in the region were reinforced by troops from several nations. The United States sent the Ninth Infantry and a regiment of marines from Manila as part of an allied expedition that tried to reach Peking. When the Chinese insurgents, who were known as the *Boxers*, badly defeated this force, several nations sent larger contingents. The Fourteenth Infantry and Reilly's Battery were part of the American force ordered to Tientsin.[2]

We embarked on a tramp freighter in Manila Bay with horses, guns, supplies, and ten thousand rounds of ammunition. I had charge of much of the loading. In order to bring our battery to war strength, other units furnished men by emptying their guard houses and marching their prisoners[3] to the dock under guard. They made up our deficiency in horses in a similar manner.

We sailed from Manila on July 15, 1900. Besides the battery, the ship carried a battalion of the Fourteenth Infantry, their mules, and their wagon train. There were no staterooms or any place to cook or serve food except the ship's galley. We had coffee just once a day, and our food consisted of canned beef, tomatoes, and hardtack distributed twice a day. All slept on deck or in the hold. Although there was little complaining, Captain Reilly sharply reprimanded an officer for some casual remarks about the conditions.

It took eighteen days to reach Taku, where we anchored August 1, among hundreds of ships, eighteen miles from the mouth of the Pei-ho River. At once, a barge was brought alongside for the guns and caissons, and a small coasting steamer was lashed to the ship to receive the animals. I had charge of unloading the animals. Because the Anamese crew refused to help or to operate the winches, we had our men operate the machinery to hoist the horses in a stall and lower them into the boat. We handled the guns the same way. During the night of August 2, a heavy gale arose, and the cables holding the barge and the steamer parted. I was on the steamer with nearly all of the horses and mules. I told the captain to take the boat to Taku, a short way up the Pei-ho, on the morning tide. We unloaded and returned on the afternoon tide. In the meantime, the barge with the guns had parted its cables and drifted to sea. I had enlisted a stowaway sailor on the barge, who had no teeth, after leaving Manila. By rigging a sail from some Korean matting on board, he got the barge under control. Otherwise, it would probably have foundered in the Gulf of Chihli, losing the battery. In this condition, a sailboat from Korea came along with an English customs officer, en route to Taku. The boat conducted the barge to the Pei-Ho River, and I met it crossing the bar as I returned.

On August 3, we unloaded the rest of the cargo and entrained at Taku for Tientsin. There, we found orders for the allied commands, which were to march late in the afternoon of August 4. The column represented many nationalities with every kind of transportation from camels to rickshaws to coolies. The American contingent consisted of the Ninth Infantry, the Fourteenth Infantry, a regiment of marines, and Reilly's Battery. The Ninth Infantry and the marines had taken part in the previously unsuccessful advance on Peking, which had resulted in heavy losses, including Colonel Liscum[4] of the Ninth Infantry.

The battery bivouacked on the road at about 10:00 P.M. About midnight, heavy firing began ahead of us. When we resumed the advance at dawn, we came on the Japanese engaged against an entrenched Chinese position near Pei-Tang. There was no unified command, and each contingent acted independently. We saw a retreating column of Chinese and fired on it a number of times. That night we bivouacked at Pei-Tang. The next day, the Russians took the lead, followed by the Japanese. When we advanced on August 6, we found ourselves marching abreast of a British column of Sikhs and Bengal cavalry. As we approached Yang-Tsun, we came under considerable fire. Our battery went into action in a field of tall *kaulin,* or corn. Because we could not see the enemy, I placed a bamboo ladder I carried on the railroad embankment not far from the battery; from this I could adjust the fire on the Chinese position. The infantry deployed, advanced, and drove the Chinese back. During this action, we saw nothing of the Russians and Japanese, who had taken different routes.

From long habit, the three platoons of the battery were inclined to act independently, but Reilly checked us by saying that the battery lacked fire discipline. At one time, General Chaffee,[5] our commanding officer, sent for me and gave me some orders as to a target in a most considerate way. I do not know why he did not send for the captain. Throughout the day, the heat was terrific. Our canteens were soon empty, and men and horses suffered horribly for water. We camped on the Pei-Ho River and, in spite of the many bodies floating in it, drank the water. Some filter bags were furnished, but we could not wait. We had already seen many bodies in the river at Tientsin and had become accustomed to it.

Lieutenant Burgess rejoined the battery at Yang-Tsun, thus completing our officers. During the next few days, we followed the Russians and Japanese. Slaughter and destruction marked their path. Along the road, heads were mounted on poles; bodies littered the roads and floated down the river. In the poor houses of the villages, women and children were slaughtered. Nothing living was left. It filled us with disgust and contempt, especially for the Russians, who were the lowest class of brutes. I saw women at one place running from them and drowning themselves in a canal. The heat was terrific, and the road was strewn with unconscious soldiers. Our horses could barely pull the guns, even with everyone dismounted.

On the night of August 13, we bivouacked on the road, not knowing our location or anything of plans. During the night, we heard a great volume of fire of artillery, machine guns, and rifles ahead of us. At the first light, we marched with my platoon leading because it was ready first. As we were following a deep, sunken road, we came on a French battery halted and blocking the way. I ordered the sections to climb the embankment. As the six horses of the leading section strained up the almost perpendicular wall, the trace of a wheel horse broke. The other wheel horse seemed to know that his time had come. With unbelievable power, he pulled the carriage up with little help from the horses in his front. After that, he was a hero in the battery. His name was Putnam, and nothing was ever too good for him. He became a legend in the artillery and died in retirement in Manila.

After advancing some miles, I saw the great walls, which we knew was Peking, and drew near the firing. I moved along the road where the wall of the Chinese City extended beyond the wall of the Tartar City, as we later knew them. There, I found great numbers of Russian dead, swollen and bloated in the sun. On reaching the great pagoda over the gate at the junction of the two cities, I found many Russian soldiers crouching in the buttresses, completely whipped. Two Russian guns inside the gate of the Chinese City were not even manned. They occupied the only spot from which I could enfilade the Tartar City wall between the two cities and cover the advance of our troops. I had my men pull them aside and at once placed my guns and began firing.

In the meantime, the Fourteenth Infantry had scaled the Chinese City wall along which I had advanced and covered me with their fire until I could go into action. Other troops entered by the east gate of the Chinese City. Afterward, we learned that the allied generals had agreed for the Americans to advance as we did and for the Russians to attack the east gate of the Tartar City a mile or more to the north. They thought that they could reach the legations by our route, so during the night they disregarded the agreement and attacked our objective, with the result that they lost heavily and failed. I continued my fire along the top of the Tartar City wall parallel to which our troops advanced to the legations, entering the Tartar City near the legations by the watergate under the wall and the Chien-Men gate, which was opened after the legations were relieved. Unfortunately, the British, who were better informed, entered the watergate ahead of our infantry. During the action, a Russian officer sent me a note asking me to cease firing until a Russian column could pass in front of my guns. I did so, but it soon abused my confidence. After several alien contingents, including these Russians, had entered the Chien-Men gate, Captain Reilly decided to blow open the first gate of the Imperial City about a hundred yards inside the Chien-Men gate. As he was ready to fire, a Russian colonel with his flag wrapped around his body placed himself at the muzzle of the gun and told Reilly not to fire as the gate belonged to the Russians. Reilly would have fired if the Russian general Linievitch[6] had not asked him to wait until he could confer with General Chaffee. Reilly consented, and Chaffee sent him orders not to fire. It appeared that Linievitch and Chaffee agreed that the Americans and the Russians would attack together the next morning and jointly capture the Imperial City.

Late in the afternoon, I sent a scout to locate the American troops, and, as resistance had practically ceased, I joined them near the Chien-Men gate and went into bivouac, protected by the buttresses of the great wall. The next morning, August 15, my platoon was again ready to march first, and I took the lead. We had not gone far when I was ordered to enter a great gate surmounted by an imposing pagoda. This was the Chien-Men gate of the Tartar City. As my platoon had had most of the action on the preceding day, I was ordered to park my guns and the caissons of the battery under a high wall about a hundred yards in front of the two massive gates. This, I soon learned, was the first wall of the Imperial City. A great deal of firing was coming from the front over our heads and from both flanks of the Tartar City wall. The other four guns were taken up ramps inside the Tartar City wall, and two were placed on each side of the pagoda firing to the front over our heads and to both flanks of the Tartar City wall. Some infantry were even placed on the wall and fired over us to the front. The Chinese fire from the front became more intense.

Very soon I received an order from Captain Reilly to blow open the gates of the wall behind which I was parked and attack. The gates were of massive hardwood about ten inches thick, covered with sheet metal and barred on the inside by beams at least ten inches thick. Looking through the crack where the gates came together, I located the bar. I placed a gun about fifteen feet from the gate, scratched with my thumbnail a cross on the metal covering of the gate opposite the bar, and ordered the gunner to fire at that spot. The gunner, who was Corporal Smith from the mountains of Tennessee, had an accurate eye. Much to my relief, the first high-explosive thorite shell cut its diameter through the gate and the bar. I had been afraid that it would burst on impact and that we would be hit by the fragments. I made two other scratches opposite the uncut parts of the bar, and, with two more shells through them, the bar was cut, and the gates swung open.

There were three of these gates, and I cut the bar of another gate at the same time. We saw in front, a few hundred yards away, another wall at least twenty feet high surmounted by a great pagoda. The wall had a crenellated parapet[7] from which heavy rifle fire was coming. I had already noted that several infantrymen who had tried to scale the wall that we had opened were shot down. There was also a good deal of rifle fire from our flanks. I ordered both guns to fire percussion shrapnel rapidly at the crenellated parapets of the wall in our front. The impact of the shrapnel scattered the marble of which the parapet was made, and the bullets of the shrapnel made a hail of lead beyond. The fire from there soon ceased. Both sides of the approach to the second wall were enclosed by high walls about a hundred yards apart with the space between them paved. We did not know what would happen if we advanced into this great court, but it was necessary to occupy the second wall with our infantry to cover the advance of the guns to it. That morning, I had seen some ladders with the Japanese, so I sent for them. Under cover of the fire of my guns, the Fourteenth Infantry advanced to the second wall, placed the ladders, and scaled and occupied it. I then advanced my guns rapidly under the protection of their fire to the second wall.

Soon after I began firing through the gates of the first wall, a soldier came and told me that the captain had been killed. I told him that it could not be true and not to say that to any of the men. Earlier wild rumors circulated, one of which was that he and I had been killed at Yang-Tsun. Shortly afterward, Captain Scriven,[8] of General Chaffee's staff, who came to ask me about some ammunition for the battery, said that Captain Reilly had been killed and that I was in command of the battery. I told him that Lieutenant Burgess was senior to me and should succeed to command. The shock was indescribable. Apparently, Captain Reilly had been standing on the parapet of the Chien-Men gate with General Chaffee on one side and Major Waller[9] of the marines on the other, watching the performance of my guns, when a bullet hit him in the mouth, killing him instantly.

At the time, however, I had much else to think of. The gates of the second wall were blown open like the first, only to find a third similar wall in our front. My guns drove the defenders from the parapet. The infantry scaled it under my fire, then used rifle fire to cover my advance to the wall. On reaching it, we blew the gates open, only to see a fourth wall flanked by towers. The infantry occupied the third wall, and I advanced under its fire to the fourth wall. As I was about to fire, Captain Crozier[10] of General Chaffee's staff gave me orders from General Chaffee to cease firing.

Now we learned more of the perfidy of the Russians. General Chaffee had ordered Captain Reilly not to fire on the first gate of the Imperial City on August 14 because the Russian general Linievitch had agreed with him to attack the Imperial City with the Americans on the fifteenth. Instead of doing so, the Russians encircled the Tartar City during the night and occupied and looted the Summer Palace twelve or fifteen miles to the west on the fifteenth while we attacked alone. General Chaffee ordered me to cease firing at the gate of the Forbidden City because of the protest of the Russians who wanted to enter it with us. Naturally, we were very bitter against the Russians. American troops, however, occupied the ground gained and guarded the approaches to the Imperial and the Forbidden Cities throughout the nine months of the subsequent occupation.

Besides Captain Reilly, a number of infantrymen had been killed. We prepared the body of our beloved captain as well as possible; his and the bodies of the others were interred in the American legation grounds on August 16.

We were told that the Chinese court and its troops had evacuated the Forbidden City on the night of August 15, and the city fell into our hands without assault. Of the two imperial flags flying over the gate towers, I gave one to General Crozier and kept one. Later, General Chaffee recommended me for one brevet for the action at Yang-Tsun and for another brevet for the storming of the gates of the Imperial City. There were no decorations then.

The battery bivouacked on the grounds of the Temple of Agriculture in the southern part of the Chinese City. The next day, we were warned that a Mohammedan division[11] was advancing from the imperial hunting park some miles south of the Chinese City on Peking. We at once assumed defensive positions, with my platoon and a company of infantry established at the southwest gate of the Chinese City. As we advanced inside the wall, we came to an execution ground that was a marsh filled with headless bodies. It was necessary to drive the guns over them. The Chinese had no prisons and, thus, beheaded people for every offense.[12]

When we reached our position, I placed the guns on the wall near the gate and prepared to cover the approaches with fire. That night, I began to swell and itch from poison ivy, to which I have always been susceptible. Peking was full of the vine. My face was so swollen that my eyes were nearly closed, I had difficulty using my arms and hands, and I was bandaged with such remedies as were available.

During the night, we heard a violent explosion inside the city wall and not far from us. At daylight, a patrol discovered an arsenal of black powder in buildings surrounded by a moat. The buildings contained thousands of seventy-five-pound baskets of powder. Had the previous night's effort to explode the powder been successful, we would have been destroyed. I reported the situation to our headquarters and received the order to burn the powder. I recalled that when the British tried to burn some powder en route to Peking, all the men engaged in the work were killed. But I used some Chinese workers to spread a basket far removed from the magazines and burn it by a powder train. The heat made it dangerous to bring any more powder to that place, so I decided to dump the powder into a moat. Each day, I had some hundreds of Chinese brought to the magazines, searched for matches and tobacco, and then had them carry the baskets to the moat. In the end, the moat was filled, but the powder was under water.

All this time, my suffering from poison ivy was horrible. Because my face was so bandaged I could scarcely see, my arms were so swollen that I had to carry them in slings, and my orderly had to feed me. In the meantime, the alarm ceased, and, with the powder destroyed, we rejoined the rest of the battery in the Temple of Agriculture.

In a few days, a great parade was organized to march through the Imperial and the Forbidden Cities. One officer, or soldier, was selected to represent each company in all of the contingents. I represented Reilly's Battery. It was an imposing spectacle, with all the world passing in review. We formed at the Chien-Men gate and marched over the route that the Fourteenth Infantry and my guns had captured on August 15. No foreigner had ever before entered the Forbidden City. We marched around the throne rooms and temples to Coal Hill, north of the Forbidden City. Although the buildings were impressive, the lotus ponds were the most beautiful.

During the occupation, American officers escorted all visitors to the Imperial and Forbidden Cities. In this way, I later saw the throne rooms and the apartments of the court and the retainers. There was richness and luxury, but all was alien to the Western mind. The gorgeous peacock throne room was magnificent. Many articles had been given by the crown heads of Europe. No looting took place, and all was left as we found it.

Without Captain Reilly, there was deep depression. We realized that a great change had taken place, though all tried to maintain his standard of efficiency. Men and horses were worn out, tired, and reflected the hardships of the campaign. Everything had been left at Tientsin not essential to existence. The battery had taken only the ammunition and the barest necessities for the men. All needed clothing and equipment necessary for camp life, especially tent shelter because the weather had turned colder. The most malignant, deep-seated boils had begun to replace the malaria and dysentery of the Philippines. I had fifteen, mostly in pairs in corresponding glands, during the next few months.

Under these conditions, I was sent to Tientsin for our baggage and equipment. At Tung-Chow, our depot in the Pei-Ho River about twenty miles from Peking, I procured a Chinese junk, towed by eight or ten coolies. We had a few rations, and the coolies gathered vegetables along the banks. My orderly cooked our food. We slept in compartments of the junk in the open. On our four-day trip to Tientsin, we would tie up at some village at about nine o'clock at night and resume travel at daylight. As we were passing a deserted village one night, a voice called from the river bank to know if Lieutenant Summerall was on the junk. I recognized the voice as that of my classmate Lieutenant John M. Palmer and told him that I was. I landed, and we had a reunion in that tragic and desolate place. Although he also went by junk to Peking, I never saw him there.

At Tientsin, I procured eighteen junks with coolies to pull them and loaded all of the battery's equipment on them. Relays of six or eight coolies in a sort of harness attached to the tow line pulled the junks, which used sails as much as possible. In the mornings, the coolies waded through ice. Progress was slow against the current. It took several days to reach Tung-Chow, which was the head of junk navigation on the Pei-Ho River. There, wagons met us, and we proceeded to Peking. We pitched tents, equipped kitchens, and mounted tarpaulins to shelter the horses, guns, and carriages. Our camp was made as comfortable as possible.

We found recreation in visiting the shops in the different concessions and buying robes, silks, cloisonnés,[13] and antiques. Although we had captured the Imperial City, the British took possession of the imperial silk-go-down, consisting of thousands of rolls of the finest silks, and sold them to us at a high price, while the Russians took the furs and sold them to us. The allied forces divided the Chinese and Tartar Cities into sections to be occupied by the Americans, the British, the Russians, the Japanese, and the French. Each national element generally kept to itself, though our battery established a friendly relation with the Twelfth British Battery from Lahore, India. The American legation also invited us to call and to tea, where we met some young ladies.

The climate of Peking is about like that of Philadelphia or New York. Snow and ice and the cold winds from the Manchurian mountains brought in a rigorous winter. Coal came in by camel trains. The

Chinese mixed powdered coal with clay to make small balls. Ignited in cylindrical braziers, they were brought into our tents when they glowed red (showing that the carbon monoxide had gone). Some were brought in too soon in other commands, and a few people were asphyxiated. We eventually received small tent stoves and burned coal.

In the meantime, the Chinese Li Hung Chiang,[14] who took the leadership in treating with the foreigners, visited our battery. He was an impressive character and embarrassed us by asking where we had gotten certain articles of Chinese furniture. Although Captain Reilly sternly forbade the taking of anything, a certain amount of looting became accepted, and our men and officers had a small share. Some officers even accumulated rich stores of treasure.

When Captain Reilly was killed, Captain Thomas Ridgway,[15] Fifth Artillery, was ordered to command the battery. He arrived about two months later. We all knew him well and liked him. He was a fine man, an able officer, and the battery functioned admirably under him. Lieutenant Harrison Hall[16] also joined. The normal routine was followed, including an officers' school, which I was detailed to teach.

During the winter, the Russians and the Japanese gave public demonstrations of firing by their field artillery in the imperial hunting park. The British and the American batteries gave a combined demonstration, firing as one command. The French not only did not fire but also would not allow anyone to approach their guns. Because the seventy-five-millimeter gun was a major secret with them, they did not associate with the other troops. The Italian component was insignificant and remained to itself. The Germans did not arrive until the spring.

My increased pay and the small cost of living enabled me to send money to support my sister's family and to pay Mr. Jones all that I had borrowed from him. Apparently, he had provided in his will for $500 for me, which covered the loan for my brother. Although I had paid this back to him, when he died in 1902 there was a bequest in his will of $500 for me. This proved to be a godsend afterward. I also sent Dr. Porter money for all he had spent on me during the time he took me as a "charity boy." Before leaving Manila, I had been examined for promotion to captain, and I expected that in the near future.[17]

Throughout this period, the most important thing in my life, and that which was uppermost in my mind and heart, was Miss Mordecai. We wrote constantly to each other. Every mail brought the most beautiful letters from her to inspire and hearten me. I now felt free to marry, and, through our letters, we determined to be married whenever I returned to the States. I carried her photograph in my watch. Early in 1900, she wrote me that she would visit friends in Manila and gave me the transport schedule both ways. Her father had been transferred to the command of Benicia Arsenal, California, which brought her much nearer to me. There were no regular mails, and letters took a long time in transit. I determined to meet her in Nagasaki during her return trip on the transport. No one had ever thought of asking for a leave in Peking, but I went to General Chaffee and asked him for three week's leave to go to Nagasaki. He was most kind in his attitude and at once told me that he would grant it. I wrote to Miss Mordecai and then sought some way to go.

Because there were no commercial or American ships available, I went to the Japanese intendance office and requested authority to travel on a Japanese ship. The officer was most cordial and insisted on my drinking several bottles of bitter Japanese beer with him. Even though I had never drunk beer and the ordeal was very difficult, I could not give offense by refusing. He finally gave me a paper and told me to take it to the captain of a certain ship off Taku, in the Gulf of Chihli. I then went by train to Tientsin and asked the quartermaster to send me out to the ship that night, which he agreed to do. At Taku, a small launch took me to the ship, anchored about eighteen miles off the mouth of the Pei-Ho River. As we approached the ship in the dark, the water was too rough for the launch to go alongside. The captain said that I must jump and catch a rope ladder on the side of the ship when the waves tossed the launch close enough. This I did and by good fortune clambered up to the deck. My bag was then hoisted aboard by a line.

I gave the captain the paper from the Japanese intendance office. His welcome was friendly, but he indicated that they had no foreign food aboard. I told him I could eat anything. I used chopsticks to eat the stew and rice served for food. We sailed soon after I arrived, and the next morning the ship anchored close in at Shanhaikuan, where the Great Wall goes into the sea. It could be seen climbing up the hills for a long distance. We then sailed for Chempulo,[18] Korea, and then to the entrance to the Inland Sea at Shimonoseki. There, I left the ship and went to the station to buy a ticket to Nagasaki, but I could not make myself understood. Happily, a Japanese sailor who was on the American gunboat stationed in Japanese waters appeared and interpreted for me.

On reaching Nagasaki, I went to the quartermaster, who had charge of coaling and supplying all transports returning from Manila, and asked when the *Sheridan* would arrive. He said that he knew nothing and appeared to resent my asking. According to the schedule sent to me by Miss Mordecai, the ship was due in two or three days. When she did not come, I cabled her in Manila. The cost was $3.85 per word, and I was brief. It did not occur to me to head my cable at Nagasaki. When there was no reply, and in the course of time, I sent other cables, without result. The quartermaster professed ignorance of the ship and did nothing to help me. While at Nagasaki I managed to see a Japanese cherry blossom festival and to make an overnight trip with a young American doctor stationed there to a famous resort at Unzen.

Finally, with a heavy heart and bitter disappointment, I had to return to Peking, which I did aboard a small Japanese passenger ship going to Taku. On board was a theatrical troop of very decent people. When we reached Tientsin, the hotel had no room for them. I went to the American missionaries, who had abundant room, and asked them to shelter the women. They at first refused, but on my insistence they consented. We were not favorably impressed by the missionaries whom we saw. The next day, I took the train for Peking. When I reached my tent, Lieutenant Burgess, my tentmate, said: "Some fellow has been cabling you from Manila. The cables are on your table." There were the replies to my cables from Nagasaki. Miss Mordecai thought I was cabling from Peking. I learned in due time that she had reached Nagasaki the day that I reached Peking. Because there was no prospect of my returning to the States, the situation seemed hopeless.

While I was away, a large force of German troops arrived under the command of Count von Waldersee.[19] They staged an imposing parade the day of my return, displaying their gaudy uniforms and the goose step. Their artillery was very impressive. It was evident that they were well trained and efficient. The firing exhibition in the hunting park showed speed and efficiency. The German soldiers were subordinate, but their officers were arrogant and patronizing. We respected their efficiency, but we did not like them, and we had nothing to do with them. Count von Waldersee wanted to assume command of all the Allied contingents, but the Americans, and perhaps some others, declined to recognize him.

It was evident that bad feeling existed between the Russians and the Japanese, and we took it for granted that they would soon be at war. Our sympathies were with the Japanese, for we hated the Russians for their perfidy, their cowardliness, and their brutality. The Japanese showed great friendliness to the British and the Americans. Their greeting on the river was to grasp three fingers of the left hand in the right hand and say: "England, Japan, and the United States." The Russians began to remove to Port Arthur all rolling stock of the railroad that they soon would seize when they went to war against Japan in 1905.

During the winter, the south wall of the Chinese City was breached and the railroad brought into the space between the Temple of Agriculture, occupied by the Americans, and the Temple of Heaven, occupied by the British. It should be explained that there was no temple in the Temple of Agriculture, which consisted of a walled area of about thirty acres with a large stone platform and some stone images. The Temple of Heaven contained a larger acreage and the real Temple of Heaven, which was one of the most beautiful buildings in the world. The Chinese had never allowed the railroad to come nearer than about three miles from the city. They regarded as desecration the laying of tracks, building a depot, and running trains into the ancient city.

During much of the time I was judge advocate of a general court-martial that tried many cases. Often, Chinese witnesses were called, and they at once kneeled to be sworn and testify. They were evidently overcome by fear. In one case, an American soldier was tried for killing a French sentinel. It looked very serious. When I interviewed the French surgeon who treated the victim, he said that the American soldier's blow on the victim's head was only a coincidence and that the Frenchman would have died anyhow from a brain trouble. I did not believe him, but, because of his testimony, the court had to acquit the accused. All minutes of the trials and the records were in longhand and required much time to compile.

Chapter 8

Back to Manila and Home

In May 1901, orders came for the battery to return to Manila. When we marched away, a large contingent of British officers, including General Gaselee,[1] the commander, rode several miles with us. Colonel Wint,[2] who commanded the column of artillery and cavalry, did not want the men to visit the Chinese towns. He camped several miles from Tientsin, thinking it too far for the men to walk to the city. He then rode off to town, followed by his second in command. During the night, many men of the battery and the cavalry went absent without leave to see the old city; all returned before morning.

When we boarded the ship off Taku, the Chinese crew refused to work. Our soldiers had to handle all of the cargo and the horses. Again, there were no accommodations for cooking or sleeping, so officers and men lived on canned goods with a little coffee and slept on deck or in the hold. The ship was so hot for the horses that we took the risk of bringing them on deck in relays to get fresh air and to cool off. At Manila, we were lightered on cascoes and returned to our old camp in the Agricultural Grounds. As there was no prospect of coming home, all were much depressed. Then, suddenly, an order came for the battery to return to San Francisco on the *Pak Ling*, a fast tea ship, as a guard for some hundred prisoners. Only the personnel were to go on board, leaving horses, guns, and equipment behind.

Captain Reilly had always said that he would not have a married lieutenant in his battery. All of us lieutenants were engaged, but his death and the prospect of promotion to be captains changed the inhibition for us. I had never mentioned Miss Mordecai to the others. As soon as the order came, I hastened to the cable office and sent her a message as to the probable date of our arrival in San Francisco.

While the *Pak Ling* was a beautiful, fast ship, she had been used only as a horse transport and had no space for passengers. We preempted empty horse stalls for our camp beds and served food on deck. The trip was comfortable. As we passed Corregidor at night and saw the lights of Manila grow fainter, I wondered how I had endured so much hardship and privation. We followed the great circle route, but we did not see the sun during the twelve-day voyage because of thick fog. Before going to China, I grew a beard, as the men had done. In China, I shaved all but a moustache. The day before reaching San Francisco, I shaved my moustache, much to the disgust of my friends. I wanted Miss Mordecai to see me as she remembered me.

Returning soldiers were a common sight in San Francisco, and no one paid attention to the landing of Reilly's Battery. As soon as we had gone into the cantonments at the Presidio on June 30, I took a train for Benicia and arrived as the Mordecai family was at dinner. She was standing on the porch at the top of the steps when I alighted from the express wagon that took me from the station. Much of the rest need not be told. That evening, I told General Mordecai that I wanted to marry his daughter, and he had little option, though he had no objection. She went to visit friends in San Francisco during the time the battery remained, and I saw much of her. Mr. Winfield Jones gave us a lovely dinner. After I informed my old friends, the young ladies whom I had known scratched me off their lists. I bought an engagement ring, which her father jestingly called a great extravagance. She fixed August 14 for the wedding, without realizing that it was the anniversary of the capture of Peking. Without consultation, Lieutenant McCloskey's fiancée in Pittsburg fixed the same day for their wedding. Lieutenant Burgess's fiancée in San Francisco fixed July 15 for their wedding. Lieutenant Hall had to wait a little longer. Thus, Reilly's lieutenants disqualified themselves for his battery.

When the battery was ordered to Fort Walla Walla, Washington, all the lieutenants, except me, went on leave. Captain Ridgway and I, with a skeleton crew of personnel, proceeded north by special train. During the first night, a terrific jolt threw us out of our berths. Our train had collided head on with a southbound passenger train at the foot of Mt. Shasta. No one on our train was hurt, but several were injured on the passenger train. The locomotives lay on their sides, with one of our cars completely telescoped into the car in front of it. Our Chinese "loot" lay scattered everywhere. Happily, our kitchen car remained upright, and our food was safe. After wrecking crews restored traffic, another locomotive continued with what was left of our special.

We found the old post of Fort Walla Walla deserted except for several hundred wild horses and a number of cowboys employed to break them. As we proceeded to organize the command, I was made quartermaster-adjutant and also held other staff positions. As always, our men responded, and our superior noncommissioned officers were invaluable. I selected a good set of quarters and bought a minimum quantity of furniture. By great good fortune, the wife of a noncommissioned officer then serving in the Philippines asked if I would employ her as cook. She was far too qualified for such work, but I gladly accepted her. She proved to be one of the greatest blessings of our lives. Her name was Margaret McInturff. She was keenly interested in my marriage and put the house in good condition for my bride.

Captain Ridgway gave me sufficient leave for my trip to Benicia and the wedding. My bride procured the license before my arrival to save time. Together we went to San Francisco for the wedding ring. We asked Mr. Kelly,[3] the chaplain at the Presidio when we were there, to perform the ceremony. The ceremony took place at noon before a flower-draped, improvised altar in the large house of the commanding officer of the arsenal. Then the reception and the lunch followed on the beautiful lawn. No one was ever so lovely as the little bride. The whole setting was like a paradise. My promotion to captain took place on July 1, but I had not received my commission. Nonetheless, I wore my captain's bars on my dress epaulettes with my full dress uniform and saber. This was the culmination of life and the richest reward that any man could have.

We traveled by train from Oakland, to Portland, and then to Walla Walla. The trip was terrifically hot, but, with the car windows open for air, we choked with dust. We kept wet handkerchiefs to our faces. The train carried freight, animals, and passengers and stopped at every station. We reached Walla Walla about eight o'clock. On arriving at our quarters, we found Margaret suffering great pain, which required my little bride to spend most of the night nursing her. Margaret recovered somewhat next day and managed to prepare our meals.

After a week, orders came for me to proceed to Fort Lawton, Seattle, Washington, and take command of the 106th Company, Coast Artillery Corps. My commission as captain arrived, and I received the munificent salary of $180 per month. Fort Lawton was a new post occupied by just two companies of coast artillery. As the senior officer, I commanded the post. We occupied a new set of captain's quarters but for a time ate on a packing box. We procured an ignorant Japanese as cook. In a few days, our good Margaret wrote that she would soon join us, which we welcomed wholeheartedly. She prepared excellent meals and kept the house clean. Because there was no other woman at the post, I took my little wife with me to inspect the guard at night, fearing to leave her at the house. As there was much to do, the winter passed quickly. Among the subjects taught to the men was cordage, which my wife enjoyed, working over the knots with me, learning their names and how to tie them. She was keenly interested in all that I did, such as planting trees to improve the post. At that time, Fort Lawton was seven miles from Seattle, though the city has subsequently grown around it.

Chapter 9

The Land of the Midnight Sun

Early in April 1902, I received a telegram to send a company of men to Skagway, Alaska, by the first boat. I was able to embark the Thirty-second Company, Coast Artillery Corps, the next day. Later, another telegram directed me immediately to take my company to Skagway. Again, I caught a boat and left on May 11. Our baby was expected in June, and I was compelled to leave my wife with only the good Margaret to care for her. The situation was heartbreaking, although there was a good contract doctor[1] stationed at the post.

On arriving at Skagway, I received orders to send the Thirty-second Company to Valdez, Alaska, which I was able to do at once. We replaced a company of colored troops of the Twenty-fifth Infantry, which returned from Skagway to the States. I soon learned that England, through Canada, had laid claim to all of southeastern Alaska.[2] As a result, the Northwest Mounted Police forced the American customs station forty miles back to the summit of the mountains twelve miles from Skagway and had occupied the town shortly before I arrived. The Canadian police drove American gold miners from their possessions on the Porcupine River and were ready to seize Skagway and Valdez and to take over southeastern Alaska by force. Secretary of War Elihu Root stopped this daring plan by occupying the disputed territory with two companies of coast artillery one thousand miles apart. Paradoxically, a Virginian, Colonel Zachary Taylor Wood, commanded the northwest police and Major Snyder of Boston commanded the station at White Horse on the Yukon, with whom we had to deal. In this situation, because I expected to be attacked, I sent a field gun to Skagway and prepared a defensive position.

In the meantime, I was desperately uneasy about my wife, who was alone except for Margaret McInturff and the contract doctor. A telegram finally came that Charles Jr.[3] was born June 16. Later, I learned that the lives of both were saved by a miracle. Dr. Wickline,[4] the post surgeon, summoned a specialist from Seattle to Fort Lawton at the critical moment. Both mother and baby were very ill, and no one who was near knew what to do. Her aloneness was heartbreaking, and she and the baby were nearer to death than life. I have never been able to reconcile myself to this separation. The army is cruel to families. When Dr. Bailey from Skagway went to Seattle for his discharge, he arranged for them to come to Skagway three weeks after the baby was born. The boat arrived after midnight, and I found them very ill in the care of Margaret. The baby was starving. We found an excellent doctor and some good women in Skagway, including a foster mother, Mrs. McCrea, who knew what to do. The leading saloonkeeper's wife and the doctor's wife were very fine women who also tried to help. The baby's life was saved, but both continued very ill.

Under the conditions, the army ordered me to go to Haines Mission on Lynn Canal, about fifteen miles south of Skagway, locate a military reservation, and begin construction of a post for a battalion of four companies of infantry. I procured a small launch, employed a surveyor, and took about twenty armed men to Haines Mission. This was the heart of the disputed territory and the beginning of the trails to the Yukon and the Porcupine River. On landing, I found a number of men of the Northwest Mounted Police. The message to me had been sent in the clear over Canadian telegraph lines as there was no American telegraph or cable line to Alaska. The northwest police forbade me to begin any construction and said that this was disputed territory. I replied that this was American territory and that, if they interfered, I would use force.

After seeing what I could from where we landed, I decided to place the post behind a high mountain to the north to protect it from the north wind and in a valley protected by a ridge to the south. I had an old order designating a military reservation said to be at a rock at Haines Mission. Selecting a rock that would serve the purpose, I directed the surveyor to begin the traverse. The reservation took in the entire area across the peninsula from Haines Mission Harbor to Pyramid Harbor, including the abandoned Indian village, and extended to take in a lake high up on the peninsula for a water supply. In all, it covered a space about two miles wide and four miles long. The northwest police followed me all of the first day, then left. The Haines peninsula extends several miles into Lynn Canal between the mouths of the Chilkat and Chilkoot Rivers. The terminal moraine of the glacier that formed the Chilkoot River extends across the upper end like a great railroad embankment. To the south is Pyramid Harbor, which is a bay formed by the cataract from the Muir Glacier, falling perhaps thousands of feet in a river of foam. The roar of the cataract is like the noise of many trains. Lynn Canal is a great fjord one hundred miles long from Juneau to Skagway, from perhaps five to ten miles wide and thousands of feet deep. Prevailing high winds make the water very rough.

During our first day in the area, one "Shellgame" Jackson, a noted claim jumper, and others protested that I was taking their lands, but I referred them to Washington. The nights were bitter cold, but we kept fairly warm in our dog tents. The entire country was covered with dense jungle and enormous timber. At places, the storms had blown down so many trees that it was difficult to get through them. I had visited the lake from which I proposed to get water by following a stream from it to Pyramid Harbor. I detailed a corporal and several men to follow the stream to the lake to take samples of the water to send to Washington for testing for impurities. I took the men to the stream, but, when I left them, they were afraid to go without me. Therefore, I accompanied them, sharing the corporal's blankets and dog tent that night. I have found that men are usually afraid of jungle or dense woods where no landmarks can be seen. We built a raft, took soundings, and collected samples of water at different depths. I did not want to use the glacial water as it had impalpable powder[5] from the erosion of the rocks. I also located each building of the proposed post.

I returned to Skagway on Saturday night. When I departed the Haines Mission dock, with about twenty men and some civilians on the launch, a heavy gale was blowing, but I felt that I must try to go to my loved ones. When we left the shelter of the high mountain to the north, the wind and a heavy sea hit the launch a terrific blow. The glass around the engine house cracked, and water came in. I saw that we would capsize or founder, and I told the captain to turn around and return to Haines Harbor. He said that we would capsize if he tried to turn. I knew we could not go on, so I ordered him to turn. By a miracle, we made the turn, and the engine did not stop as it had been doing. All felt that we had narrowly escaped. Fortunately, a large ship put in during the night and took us to Skagway. Shipping men looked on those waters as extremely hazardous, with several ships lost between Seattle and Skagway.

During that season when daylight was continuous, we had great difficulty in sleeping, for it was impossible to shut out the light. In addition, the weather was cold. We wore heavy uniforms with shirt collars turned up to keep the back of the neck and head warm. The barracks and our rooms were heated by stoves. I had two rooms at the Pullen House, kept by a Mrs. Pullen, who had gone to Alaska with the gold rush in 1897, with her two young sons and a little daughter. Because there was no suitable building in Skagway for a barrack, we had a man build an excellent house for the men to keep them warm in winter. We paid him rent. There was no town government, and our sentinels along the streets kept order. Large saloons, gambling houses, and other disreputable places did an enormous business with the prospectors and miners. The conduct of my men was without a fault. They realized their responsibility. My two young lieutenants were a great help.

While in my office one afternoon, I heard a violent explosion, and the alarm bell rang. Before I could reach the street, the guard had rushed to the scene of the trouble. I found our sentinels posted around the bank and post office. The fronts of both buildings were blown out into the street. The bank was a complete wreck, with gold dust and gold and silver coins scattered in the debris. The dead body of a large man was on the floor by a hole blown into the floor. An official of the bank came in from the back of the building, very much shaken. I asked what had happened. He said that he was counting the money to close the bank for the day when a man appeared at the teller's window holding a stick of dynamite in one hand. The man said: "Give me twenty thousand dollars, or I will blow up the bank." The teller shouted, ran, and the explosion occurred. A revolver with one chamber fired was by the body, and a stick of dynamite was in the man's coat pocket. We learned later that two sticks of dynamite had been stolen that day from the railroad magazine. Apparently, the robber held a stick of dynamite in one hand when he fired at the teller, and the concussion exploded the dynamite, killing him. I immediately placed sentries over the bank until locksmiths could come from Seattle a week later and repair the vault doors. When we left Skagway, the manager of the bank told me that they gave the sentinels a bottle of whiskey every night. The first sergeant told me that the men did not drink it while on duty.

Winter would set in early in October, with snow then preventing work at Haines Mission. Thus, I advertised in the Juneau papers for bids to clear the timber from the new post even though I had no funds for this purpose. I had to act with the greatest speed on my own authority without waiting for approval from Washington. In the end, I was compelled to pay for the cost of advertising because it was not first approved by the War Department. Jack Dalton,[6] a famous character who had built a trail from Haines Mission to the Yukon, won the contract.

The War Department inspector general appeared one day and stayed several weeks. I had no idea what he was doing. After he left, I learned that he was looking for irregularities in disposing of coal and in renting buildings, which were found to have existed in a previous command. He found that they existed at Nome, and the department headquarters wanted me to go there as a member of a general court-martial, but I was not detailed because of the condition of my family. I traveled by train to White Horse on the Yukon River and called on Major Snyder of the northwest police. Neither he nor the governor of the Yukon Territory returned the calls I made. One day, the superintendent of the White Pass and Yukon Railroad took Mrs. Summerall and me in his private car as far as Lake Bennett. When we returned on the southbound train, which we met there, we rode in the locomotive. Mrs. Summerall sat on the fireman's seat and could look down thousands of feet into the ravines of the mountains from the track that was laid on the precipitous sides of the mountains. Snow sheds covered much of the distance, but, even so, snow often blocked the trains in winter.

No account of our life in Skagway would be complete without the effect on our minds of incidents that had occurred in this unique place. Once, an epidemic of spinal meningitis took the lives of many people, seeming to kill them within minutes or hours after the attack. We were advised to wear our woolen shirt collars turned up to keep the back of the head and neck warm as a precaution.

Many desperate characters and outlaws had preyed on the gold seekers in Alaska, and murder had become quite common. The story of "Soapy" Smith[7] illustrated the conditions. He was a noted outlaw who murdered and robbed until the citizens lured him out in the Sylvester dock one night and killed him. There were also stories of shipping tragedies, of which the *Islander* was outstanding. She left Skagway one night with a full passenger load, including the family of the governor of Yukon Territory. Early next morning, two men scrambled ashore on Douglas Island off Juneau to report that, before daylight, the ship struck something and sank immediately. Because they were on deck, they could swim ashore. No others survived, and no trace of the ship was ever found. Several ships running between Seattle and Skagway when we were there were lost in a short time. The inland passage was very deep, very narrow, and often fog prevailed. Wrangell Narrows was especially dreaded by navigators and could be traversed only at slack tide.

Complaints about my taking squatter's claims for the post at Haines Mission reached Washington, which ordered General Randall,[8] the commander of the Northwest Department headquartered at Vancouver Barracks, Washington, to Skagway to investigate. He arrived early in September, accompanied by Colonel Wilds P. Richardson,[9] who had spent much of his service in Alaska. When I took General Randall to the site of the new reservation, he approved of what I had done and so reported to the War Department. Colonel Richardson, whom I liked, had been the tactical officer of "A" Company at West Point when I was the cadet first captain. He asked me if I would be willing to have him relieve me at Skagway so that he could build the new post at Haines Mission. I eagerly consented as my wife and baby were very ill and I was afraid that I would lose them if I remained. Accordingly, a battalion of the Eighth Infantry, just from the Philippines, was ordered to Skagway, and my 106th Company, Coast Artillery Corps, was ordered to Fort Flagler, Washington.

Chapter 10

The Coast Defenses

We reserved transportation on the steamer to Seattle for September 22, as the day the ship had brought an infantry battalion to Alaska. The weather was cold with high winds and sleet. When we reached the ship at night, we found that the train had also brought the riff-raff from Dawson who had left the Yukon on the last steamer of the season. They had taken all of the accommodations that I had engaged for the men. As they were mostly women from the dance halls, our troops had to occupy the ship's saloon. The men, however, enjoyed the trip, dancing and talking to the girls. The wind blew a gale nearly all the way, and we encountered very heavy seas in Queen Charlotte Sound, the stretch of water between British Columbia and Vancouver Island. Again, we saw the hand of Providence working in mysterious ways. On board, we met a Mr. and Mrs. Hawkins, who were very kind to us and became our lifelong friends.[1] Mr. Hawkins, the construction engineer of the White Pass Railroad from Skagway to White Horse and of the Dawson Mines Railroad, was returning home to Seattle with his lovely wife. Mrs. Hawkins saw at once that our baby was starving and told us of her experience with her first baby, who also nearly starved. She told us how to feed our baby, and, when we reached Fort Flagler, we adopted the brand of condensed milk that she recommended. Our baby improved, but it was years before he became healthy. My wife also grew better at Fort Flagler with better living conditions, though the winter was very severe.

The voyage to Port Townsend on the west side of the entrance to Puget Sound took four days. Fort Flagler was located opposite Port Townsend on the northern tip of Marrowstone Island, which lies on the western side of the narrows connecting Puget Sound with the Straits of San Juan de Fuca. The company and our baggage were put off at the Fort Flagler dock, while I went on to Port Townsend to secure a room for Mrs. Summerall, the baby, and our good, faithful Margaret. I returned by a post boat to Fort Flagler, unpacked the furniture, and prepared the house. Then I went back to Port Townsend and brought my wife and baby to our quarters. The people at the post did not take us in. The house was heated by coal stoves, and the post was entirely isolated. Our good Margaret's husband soon returned from the Philippines, and she left us to go with him. Unhappily, she was soon stricken with some disease and died. She loved our baby as her own, and her help to us made her the best friend we ever had. We advertised in the Seattle paper for a cook and secured a very good one. Nevertheless, our life at Fort Flagler was very rugged, and my wife was ill all the time. The water supply consisted of brackish, pumped water for the bathrooms and cistern water for drinking.

This was the first time that my company had been assigned to its legitimate duty of coast defense since it was organized at Fort Lawton. The men lived in a temporary barracks and had no stock in the post exchange and no money to buy a share. Consequently, they lived on the ration of twenty-five cents a day. The other companies at the post, who had large post exchange dividends, called us the *Hungry Sixth*. But, by good management, we lived fairly well and had high morale. Our men excelled in the post's athletic competitions.

The armament of the 106th Company consisted of a battery of two ten-inch guns on disappearing mounts and a battery of five-inch rapid-fire guns on balanced pillar mounts with a depression base Lewis range finder in a tower with other necessary plotting equipment. We at once engaged in intensive training, and the men were soon proficient in gunnery, vessel tracking, range finding, plotting, and the service of the guns. I procured reports from my friends elsewhere as to the details of firing and target practice. There were two other companies at the post, but the captains had no experience.

The commanding officer[2] was a colonel badly wounded in the Civil War and really incapacitated. Some years before, he had relieved me at Fort Lawton and had commanded the battalion of field artillery in which my battery served at the Presidio of San Francisco. His hobby was improving and policing the post; thus, fatigue duty was very heavy. When the captains told him that they wanted to hold a target practice, he stated that he knew nothing about modern armament, proposed not to learn, and would have no target practice while he was in command. A few months later, the army transferred him, and I assumed temporary command. At once, I prepared a schedule of target practice and procured a tug to tow the targets. Reasons of economy compelled us to use reduced charges of powder in a cylindrical bag supported on sticks in the chamber so as to have the charge in the axis of the gun. Naturally, the ballistics were not good. While none of us made high records, we learned much about the computation of range tables and the technique of firing at a moving target. General Funston,[3] the department commander, came to see the practice and spent some days as our guest. He was most agreeable.

Among other assignments, I was the recorder for a board to develop a water supply for the Puget Sound forts, which also included Forts Worden and Casey. We found that Snow Creek in the mountains would be ideal for a source, but the cost of piping water for twenty miles was prohibitive. During the time that I was in temporary command, I received an order from the War Department to send an officer to San Juan Island in the Straits of San Juan de Fuca to investigate the condition of the graves of American soldiers buried there. When England had determined to seize the area to the northern boundary of California, the commanding general at San Francisco had ordered Captain (afterward General) Pickett[4] and his company to occupy the forty-ninth parallel on San Juan Island, which had been originally agreed on as the boundary.[5] The British also landed troops on the north side of the island. Then the fleets of both countries anchored off the island. The American general and admiral stood firm and England eventually abided by the treaty which fixed the forty-ninth parallel as the boundary between the United States and British Columbia. During this occupation, some American soldiers died and were buried on the island. Their graves were in good condition.

Later, another incident of much interest occurred. I went to Fort Stevens at the mouth of the Columbia River, opposite Astoria, to act as counsel for Lieutenant Abbott,[6] who had been in my company at Fort Lawton, and who was being tried by a general court-martial. On arriving, I found that a fire bug had set fire to a number of buildings and that the post had been threatened with destruction. A number of suspects were placed in the guardhouse, and a detective from Portland in the guise of a recruit had been confined with them. He discovered the culprit, who was transferred to a magazine in one of the batteries. The magazine could be entered only by passing through two successive iron gates that were securely locked on the outside. Lights were kept burning, and the officer of the day was required to make frequent inspections to verify the presence of the prisoner. Lieutenant Abbott, on one of his inspections, discovered that both iron gates were open and the prisoner, who had unscrewed the nuts on the inside that held the bolts of the locks, was gone. Lieutenant Abbott was immediately placed in arrest and charges preferred for neglect of duty. In the meantime, when the escape of the prisoner was discovered, the commanding officer and the superintendent of the railroad boarded a train to go to Portland to secure detectives to find the criminal. En route, the car in which these gentlemen were riding had a hot box, and the train stopped. They looked under the car to see the hot box and found the culprit riding on a brake beam. He was returned to Fort Stevens, and the bolts holding the nuts of the locks were upset so that the nuts could not be removed. When the court met, I entered a plea of not guilty on the part of Lieutenant Abbott and claimed that the conditions that enabled the prisoner to escape were the responsibility of the command and not of the officer of the day, who had performed full duty in visiting the magazines. Of course, he was acquitted. I have had a great deal of court-martial duty both as a judge advocate and as a counsel for the accused, and I have tried to do my full duty in each case. It has appealed to me strongly, and, in later years, I reviewed many cases. Military justice is one of the most serious responsibilities of an officer.

During this time, I received a telegram from the War Department asking if I was willing to accept a detail at West Point. My family was in no condition to travel, and I had no money to pay for the travel. At that time, the government did not transport an officer's family, and the baggage allowance was very small. Moreover, I was deeply interested in the work and much attached to the men who had been so loyal to me under difficult conditions. I declined the detail.

Chapter 11

The Caissons Go Rolling Along

Early in August 1903, I received a telegraphic order to assume command of the Third Battery Field Artillery at Chickamauga Park, Georgia, and march it to Fort Myer, Virginia. We shipped our furniture around Cape Horn to Philadelphia but went ourselves by railroad stateroom across the continent because of my sick wife and baby. The legacy of $500 left me by Mr. Winfield Jones was all that made this possible. Our living and the money that I sent my sister every month prevented us from saving much. We left Fort Flagler August 8. The entire company came to the dock to say good-bye. It was a sad leave-taking. The trip over the Great Northern Railroad was very rough and fatiguing, and soon my wife, our baby, and I were very carsick. By having a stateroom, we could prepare and warm the baby's food over an alcohol lamp. We changed cars on the Great Northern Railroad at Minneapolis, and the lights and the confusion frightened the baby. The journey to Chicago was particularly rough. We continued by the Baltimore and Ohio Railroad to Washington, where we were met by Mrs. Summerall's father. After a week on the road, it was refreshing to go by the trolley car to Sandy Springs, Maryland, where Mrs. Miley, Mrs. Summerall's sister, and her family were spending the summer. A room had been reserved for my wife and the baby. I left next day for Chickamauga and took command of the battery at Camp George H. Thomas.

The battery had been commanded by Captain Eli D. Hoyle, recently promoted to major. He was a superb and distinguished officer, and his men had a very high morale.[1] The Seventh Cavalry was also at the camp, and the two organizations were on the best of terms. My first duty was to take the battery across the Tennessee River and conduct target practice at a place that had been the scene of a fight between Wheeler's Cavalry[2] and Union troops in the Civil War. The march enabled me to become identified with the men and to accustom them to my methods in the field. It would have been difficult for them to accept anyone after the dominant character of Major Hoyle, to whom they were devoted. Two or three sergeants who were chiefs of sections failed to respond properly, but the leading noncommissioned officers and the men soon became loyal and cooperative. First Sergeant Hart, Stable Sergeant Vite, and Sergeant Kelly[3] were outstanding in ability, character, and loyalty. Our target practice was not very good as there had been little training for it, but the effect was to arouse interest and a desire to improve.

On September 8, the battery marched out of Camp Thomas. The entire Seventh Cavalry lined the road, and the band played "Garry Owen" and other familiar airs. It was a great tribute and stirred the battery with pride and regret at parting from their campmates and friends. The battery with its train made a long column and presented an imposing spectacle. I had three very capable lieutenants who were assigned to the necessary duties for the march. The distance was about four hundred miles, one of the longest marches ever made by field artillery. During my service in the Philippines and China, I had learned much about devices and methods to protect the horses from harness sores and to keep them within their endurance. The battery was well prepared, and I had full confidence as to the results. The discipline of the men was excellent. The itinerary provided for average daily marches of about eighteen miles, depending on water and campsites. We followed this routine with little variation. The weather turned cool at once, and, when we reached the mountains, it was cold.

Our quartermaster always preceded the battery and marked the route. He secured campsites, bought forage, and made all arrangements for the night in advance of the battery's arrival. He then met me and conducted me to the campsite. Without halting the battery, we formed the park in lines of carriages and trains successively in the most orderly manner. I announced where the tents of the officers and men were to be pitched, where the kitchens were to be placed, where the latrines were to be dug, and where water could be found. The troops then dismounted, stretched picket lines, and unhitched and unharnessed the horses, and separate details prepared different elements of the camp. In a very short time, all would be in order. The horses were rubbed down and given hay. We tried to make camp by noon, have lunch at once, and afterward water and groom the horses. After they cleaned carriages and harness, the men were free to rest or to go to the nearby town until stable call at about four o'clock. After stables, supper was served, and retreat followed. The men were then free, except the guard, which was mounted and posted after retreat. Tattoo sounded at nine and taps at 9:30, when an inspection was made to see that all had gone to bed. We had reveille at 4:30 and always breakfasted, struck camp, harnessed and hitched, and marched within an hour. The campsite was perfectly policed.

Many visitors came to the camp. At one place, a large number of young girls from a boarding school swarmed into the camp and quite excited the men, but the young ladies were treated with the greatest courtesy. Throughout the Valley of Virginia, the inhabitants received us with the greatest formality and courtesy. We were the first Union troops in the valley since the Civil War, and the people wanted to show us that, as they had been valiant foes, so they could be hospitable friends. The leading citizens and city officials welcomed us on arrival and offered any service to make our stay pleasant. After entering Virginia at Abingdon, a distinguished-looking gentleman, who had been a colonel in the Confederate army, Colonel Hurd, met me at his gate. The quartermaster had arranged for us to camp in the beautiful park in front of his stately home. I demurred and told him that the horses and carriages would ruin his lawn. He was quite resentful and said with great dignity: "Sir, if General Pleasonton[4] camped here without my permission, you can camp here with my permission." Needless to say, I accepted, for we had made a very long march because a campsite could not be found in Tennessee and all were very tired.

Another Confederate, Colonel Robinson, looking the part and riding a beautiful horse, entered the camp.[5] He introduced himself and spoke with the greatest deference, using *sir* frequently. He invited me on behalf of himself and his wife to supper. I thanked him and explained that the hour was late and I had much to do in arranging for the next day's march. He replied that any hour would do, so I named a time much later. He appeared promptly and accompanied me some miles into the country, where we came to a large gate and an avenue leading to a lighted house in the distance. When we reached the door of a beautiful old home, a Negro took the horses, and we were received by the most exquisite lady, the colonel's wife. They served a most delicious supper. All the while, the colonel had been regaling me as eagerly as a boy about his experiences in the war and his being on the staff of Jeb Stuart. Apparently, he was enjoying the opportunity to talk to an officer about the subject dear to his heart. At length, I took my leave, and he accompanied my return. Just before reaching the camp, he said very deferentially: "By the way, Sir, what state do you come from?" I replied: "I come from Florida." His whole manner changed, he slumped in his saddle, and his face clearly showed that he felt that he had been imposed on. Then he said in the most disgusted and familiar way: "O hell, I thought you were a damned Yankee, or I would not have been half as nice to you." I explained that every officer and man in the battery came from the South. He countered that we wore the Yankee uniform and looked and acted like Yankees. However, we parted good friends, and no more *sirs* were used.

Our most interesting experience was at Lexington, Virginia. Before we arrived, we were met by Lieutenant Strother,[6] the military instructor at the Virginia Military Institute (VMI), and several officers of the institute, who had been sent by General Scott Shipp,[7] the superintendent, to welcome us. Conducted to an attractive campsite in Lexington, we were greeted by Judge Letcher[8] and other prominent citizens. General Shipp invited the officers to dinner at his quarters and the men to supper in the cadet mess. Many people, including ladies, visited the camp. The officers were driven about the city, and the famous mint juleps were served at some homes, including Judge Letcher's, where we called. I was compelled to decline, as it had always been my rule not to take alcohol. This proved very embarrassing and perhaps aroused some resentment. That evening before dinner, with General Shipp, mint juleps were again passed, and, when General Shipp saw an untouched glass on the tray that was being removed, he exclaimed: "Who did not drink my mint julep?" We toured VMI and Washington and Lee University, where we saw the recumbent statue of General Lee. The Jackson statue at VMI and its significance to the Corps of Cadets was equally impressive. Our fine-looking, hardy soldiers could not fail to arouse the admiration and respect of all who saw them. The only case of trespassing occurred when we left Lexington. A soldier entered an apple orchard and took some apples. I required him to walk the rest of the way to Fort Myer. Such conduct would have given the battery a bad name and, if extensive, would have cost the orchard owners much money.

We camped at New Market and went over the battlefield where

the VMI cadets had fought so gallantly in 1864. We were received with much ceremony at Staunton, where the city officials and the faculty of the Staunton Military Academy extended many courtesies. At Winchester, we heard much more of the famous Valley Campaign. When we reached Harper's Ferry, the city officials conducted us to a beautiful campsite on a precipitous bluff on the Potomac River. I objected because the traces of tent ditches covered the ground and I was opposed for sanitary reasons to camping where other troops had recently bivouacked. The mayor assured me that no troops had camped there since the Civil War. I later found that many Civil War campsites and earthworks in northern Virginia were as well marked by trenches and ditches as through they had been made recently.

We reached Fort Myer on October 20, 1903, and were met by Captain Foote[9] and his battery, who escorted us into the post. This battery and ours were to form a battalion under Major Hoyle. As we ascended the hill at Fort Myer, my precious wife and baby stood waiting for me. I could not describe my joy at seeing them. The battery made park with every man and horse with which we had marched out of Camp Thomas, all in good condition, after a march of forty-three days and more than four hundred miles over mountains and difficult roads. We occupied new barracks and stables, which seemed luxurious to the men who had lived in camp for years at Chickamauga and in Cuba. After caring for the horses and housing the men, I went to Captain Foote's quarters, where my dear wife and baby were staying. The next day, we moved into our own quarters with the furniture that had come around Cape Horn little the worse for the trip. We found good servants and a nurse, and our lives seemed luxurious after the vicissitudes of the Pacific Coast and Alaska. My wife's father and stepmother, Mrs. Miley, with her children lived in Washington, and she had other relatives and friends. She and the baby improved, and we were very happy in our home and I in my work.

We had no company fund and had to live on the ration allowance, which was far too small. By selling manure, grain sacks, and salvage, we derived a small revenue. As we were compelled to buy into the post exchange on credit, we received no dividends. There was much to do in conditioning the materiel and the horses and in training the battery in drill regulations and gunnery. On Friday afternoons, the battery was trained to execute the wild drills in the riding hall for the public. We put the guns, caissons, harness, and horses in show condition. A keen rivalry arose between the two batteries. It was necessary to keep a caisson varnished and draped and six handsome horses with shining harness at all times for the many funerals that took place in Arlington National Cemetery. Often, we fired salutes for generals and admirals. The batteries were also escorts for large funerals in Washington, such as that of the USS *Maine* dead, which were returned from Cuba, and those for Admiral Robley D. "Fighting Bob" Evans[10] and other officers of superior rank. At the inauguration of President Theodore Roosevelt in 1905, the appearance of the battery aroused much admiration. After a competition, the battery also was selected to represent the field artillery at the horse show in Madison Square Garden in New York. The men took part in the athletic competitions at the post and excelled in the tug-of-war and running.

After we received the new model three-inch counterrecoil guns, I was ordered to Rock Island Arsenal for instruction in the materiel. At the same time, new drill regulations were issued, and they required special training. Both officers and men became proficient in gunnery, marching, reconnaissance, and the occupation of position, determining firing data, and the observation of fire. During the maneuvers of 1904 and 1905, the two batteries marched separately to Mount Gretna, Pennsylvania, for target practice. The record of the firing was very creditable, and the march discipline and care of animals and equipment showed efficient field service and training. During one of these marches, the battery went to camp at York, Pennsylvania, for a week during a country fair. On the first night there, several of my men were put in jail. I sent an officer to attend the trial the next morning. He reported that the Democratic congressman in the Republican district had us ordered there and the Republican town government showed their resentment by arresting the men. When a group of the town authorities came to my tent, I ordered them out. I had "Boots and Saddles" sounded, and within an hour the battery had marched away.

The other troops at the post were a squadron of the Seventh Cavalry who had been with the battery at Camp Thomas, and we renewed our old comradeship. At one time, the army had maneuvers near Thoroughfare Gap, where the battery showed combat and field efficiency. During the maneuvers, a captain of infantry who was none other than the brutal first classman at West Point who abused his authority in reporting me at my first cadet artillery drill, tried to interfere with my watering my horses. I met him on equal terms.[11] In spite of the saloons near the post and the temptations in Washington, there was little drinking or absence without leave. The morale of the battery was high, and men felt a special pride in belonging to it. This was shown in the care of barracks, materiel, and horses and in the dress and soldierly deportment of the men.

The lieutenants, Nones[12] and Kilbourne,[13] were capable and loyal men. First Sergeant Hart, Stable Sergeant Vite, and Quartermaster Sergeant Kelly were among the best noncommissioned officers that I have known. During the target practice of Captain Foote's battery at Mount Gretna, a fragment of shrapnel hit and badly wounded one of his lieutenants. At the court of inquiry, Captain Foote asked me to serve as his counsel. The evidence exonerated him from any blame. But there arose an unfriendly feeling among his officers toward Captain Foote and me because I had objected to a member of the court who was no doubt prejudiced against him. His safety methods at target practice were not as good as ours.

I continued to help my sister's family, and we were compelled financially to forgo any social entertainments. On our baby's second birthday, I had him ride alone on my horse, and he took delight in leading the large horses at the battery stables. His love of horses and his fine horsemanship must have begun then. On the whole, this was a very happy period in our lives. We were invited to White House receptions, where Mrs. Theodore Roosevelt made us feel welcome. Much emphasis was placed on the uniform, and I was compelled to buy the new full dress and the social dress uniforms. The battery was issued an overseas cap for experiment. It was discarded. Box spurs were worn with the full dress and the social dress. Boots and breaches were worn till evening. For civilian clothes, I bought a Prince Albert coat and a silk hat, which our child called my *coachmen's hat*.

Chapter 12

To West Point

Without warning, during the summer of 1905, I was ordered to duty as senior instructor of artillery tactics at West Point. I reported there on August 21 and was able to choose an excellent house at the south end of the post next to the old Kinsley House. My family joined me a week later. When I left them at Fort Myer, they stayed with a neighbor. The baby fell out of bed and hurt his arm badly. He looked very pitiful with his arm in a sling when I took him off the train. Again, I was able to make the move only on the remainder of the $500 that Mr. Winfield Jones had left me in his will.

This began the most difficult and disagreeable part of my service, although my relations with the cadets were all that could be desired. My duties involved the personal instruction of cadets in field artillery and supervising all instruction in coast artillery. I had no assistant and depended on officers in the academic departments to instruct in coast artillery and assist in field artillery. The training was practical with the guns, horses, and instruments but theoretical in the classrooms. The equipment consisted of materiel and horses for a mobile battery, a battery of field guns for standing gun drill, mules and guns for a mountain battery, and some obsolete coast artillery guns in a seacoast battery at Trophy Point. All materiel, harness, and horses were in poor condition.

The artillery detachment of about forty-five men not only cared for the horses and guns but also did much work on the polo field.[1] The men lived in a poor building at the north end of the post, the horses in a temporary building near the polo field. The detachment had no company fund, and the dividends from the post exchange were retained to pay the assessment in joining. The food was insufficient, with a ration allowance of twenty-five cents a day. The routine of the men was wholly unmilitary, in part because they had been regarded and treated as laborers; desertions were frequent. All conditions were the reverse of the beautiful battery I had left at Fort Myer.

I had never known the commandant of cadets,[2] who was a cavalryman, and his assistant,[3] who was an infantryman. Both, I soon learned, were hostile to me and said derogatory things about the artillery to others. Bad as conditions were, I began at once with cadet training. I found the cadets responsive and attentive, and, because they were ignorant of the deplorable condition of the horses and materiel, they were not deterred by it. For some reason, they at once gave me the nickname of "Honest John," which name I have always regarded as a compliment. I proceeded to make soldiers of the men and to place the materiel in creditable condition. I induced the post exchange council to retain only 5 percent of our share of the dividends, thus improving the mess. I ordered an inspection on Saturday morning, at which the men appeared badly clothed, unshaven, and dirty. I explained that they must comply with my orders to appear properly, but I did not punish them. The next Saturday, when they were no better, I gave some extra fatigue to a few. Five men deserted. By talking to them, appealing to their pride, and treating them as soldiers, I gradually built up an excellent standard in soldierly dress and deportment, care of barracks, and condition of horses and materiel.

Eventually, we held spirited and fast mounted drills on the plain with both soldiers and cadets as drivers. Cadets were always the cannoneers. The noncommissioned officers became efficient as instructors of cadets in the service of the guns and gunnery. We improvised plotting boards and taught fire control, the preparation of firing data, and the service of the fixed defense guns at a moving target represented by a launch. By the summer of 1906, I was able to take the first class to Fort H. G. Wright on Fisher's Island for target practice with the different caliber of guns. The field artillery practice was held at West Point, with targets on Crow Nest.

The new military academy construction program inaugurated about that time included large barracks and stables at the south end of the post for the field artillery. On completion of these we moved, immeasurably enhancing the pride of our men. In the meantime, through friends at the War Department, I was able to equip the coast artillery with the latest-model plotting equipment, construct base end stations with telescopes, and install searchlights for night exercises. The department furnished a battery of six-inch guns on disappearing carriages at Trophy Point and a battery of four twelve-inch mortars in Execution Hollow. At my urging, the Board of Visitors got money for houses for my married noncommissioned officers; fourteen of whom became captains in World War I. When the number of men and horses increased to those of a service light artillery battery, we added a pack train to the Artillery Department. The men now became the equals of the best service batteries. Because the barracks and stables were beautifully kept, the superintendent brought distinguished visitors to see them. The mess was excellent, desertions, drunkenness, and absent without leave rarely occurred. Bowling alleys, a pool room, a piano, and a gramophone enhanced recreational opportunities. The printed menus for dinners on special days reflected pride and efficiency. I bought cows and had an abundance of milk. A garden produced our vegetables, first at the north end of the reservation, then at the Dasori Place, where we kept the cows. Carloads of cabbage from Long Island were bought for sauerkraut. In the woods behind the barracks, we even had chickens, which provided fresh eggs.

A soldier whom I had punished wrote an anonymous letter to the superintendent complaining of my using the company fund for matting and salt holders and ended by saying: "But we have the best battery and mess in the army." When Lord Kitchener[4] visited the plant, he did not believe that men lived in the barracks as they were in such a show condition with polished floors, rubber matting, mirrors, etc. The stable sergeant decorated the heel posts with straw and piled the stables with thick beds for exhibition. With the company fund, I bought the first cylinder salt holders in the army for the stalls. The pack train, with its superb mules, well-set-up *aparejos* [pack saddles], picturesque equipment of the packers, and their dashing dress of the plains, particularly impressed Lord Kitchner. President Woodrow Wilson of Princeton University made me feel that the battery was one of the most interesting things he had ever seen. Only the Japanese general Kuroki[5] and his staff showed indifference, possibly because they were inflated over the victory in Manchuria.

The dashing graduation exhibition drills delighted cadets and visitors alike. During the summer, we organized the cadets into a battalion of field and mountain guns capable of adjusting and concentrating the fire of all batteries on targets at Crow Nest. On weekends, I took a battery of the first class on Friday afternoons for a practice march and a problem. Then we made camp. After the horses and guns were cared for, the cadets had supper, followed by a campfire, around which we sat and talked artillery.

Each summer, I took the first class to Fort H. G. Wright, Fisher's Island, which is off Long Island, New York, or to Fort Hancock at Sandy Hook, New Jersey, for target practice with the five-inch, six-inch, ten-inch, and twelve-inch guns and twelve-inch mortars using the fire control equipment of the fort. The cadets needed only a day or two to become familiar with the materiel, at which point we began practicing on targets towed by a tug. The records of the cadets compared favorably with those of the coast artillery. Shipping and fog often interfered and caused long waits for the brief periods required for a string of shots, fired at thirty-second intervals. One day at Sandy Hook while waiting for rain to stop, I heard a cadet call from his tent: "Now you fellows come off that. That's running it on Honest John." During the entire time, I had the greatest loyalty and cooperation from the Corps of Cadets. The trips required us to leave West Point early on Sunday morning, arriving at Fort Wright or Fort Hancock late in the afternoon, when we went into camp. We had lunch on the boat. We would return on the following Saturday.

A number of the cadets whom I taught during this tour of duty became division, corps, and army commanders or chiefs of branches in World War II, including Patton, Arnold, Spaatz, Devers, Patch, Simpson, Eichelberger, Buckner, and many others.[6] During World War II, General Patton wrote me that he was ordered to attack an almost impregnable position in North Africa. Then he said: "But I knew my artillery. I tore them to pieces with my 155s and took the position without loss." Another time, he made a breach in the wall of a fort through which his men rushed and overcame the garrison.

Each summer, the Corps of Cadets conducted combined arms exercises with all arms in the country around West Point, ending with a week of daily field problems. In one of the problems, the commandant of cadets gave me an impossible assignment and then criticized me severely at the critique. No doubt, he deliberately intended to humiliate me, with the result that I asked to be relieved. I went so far as to prepare my request for relief but then resolved that I would fight it out with him. The next day, a cadet wrote me a letter saying that he knew it was irregular but that he felt compelled to tell me that the entire corps resented the attack and supported me. In World War I, that same commandant[7] asked me to help him get a division, which I did, and, after the war, he asked me to help him to be promoted to major general, which I did.

Our little son had a great deal of sickness, including measles, earache, and indigestion. Besides the excellent doctors at West Point, we were immensely helped by Dr. Gleason at Newburgh. My wife and child enjoyed the parades, and she and I went to some of the officers' hops in Cullum Hall. We also visited our dear friends the Henry MacKays[8] on Long Island and invariably went with them to Coney Island. The MacKay girls also visited us, and two of them married army officers. At a graduation exercise, the MacKay family was with us when Captain W. R. Smith,[9] afterward major general, entered the seats. I told Mr. MacKay that the cadets called him "the friend of the underdog" because he spent his time helping the "backward man." This so pleased Mr. MacKay that he wrote a story in the *New York Times* that said: "How much better the world would be if there were more friends of the underdog."

When we occupied the new barracks, some of the steam pipes opened when the heat was turned on. I reported this to the constructing quartermaster for repair. He at once attacked me in the most vicious way for neglect in using the furnace. I requested a board, which actually fixed the fault on him. At another time, the adjutant called on me to furnish over thirty horses at night for a riding party. I declined and told him that my men and horses were too tired after work and drills to be out half the night for pleasure. He resented this and soon charged me with neglect because a horse stepped on the ice in the corral and was injured. I presented the matter to the superintendent in the adjutant's presence and had no further trouble with him. Not long afterward, he was relieved.

At one graduation, Colonel Garland N. Whistler,[10] who commanded Fort H. G. Wright, brought his mine planter[11] and enabled me to give a night drill with the searchlight and base end stations, using the mine planter as a moving target. The commandant would not include my estimate for the tactical department in the budget. At one of the exhibition drills, I talked to the chairman of the Board of Visitors, who told the superintendent to add my estimate to the budget. The superintendents, General Albert L. Mills,[12] General Hugh L. Scott,[13] and General Thomas H. Barry,[14] showed me every consideration and were invariably helpful. Between the departure of Colonel Sibley[15] as commandant of cadets and the arrival of Colonel Sladen,[16] I acted as commandant for about a month. I made some improvements in the administration of the office by appointing an adjutant. A cadet was reported for being under the influence of liquor at retreat. I preferred charges against him. General Barry said that he would not try him. He had me assemble the first class to which the cadet belonged and paroled the cadet to the class for good conduct. Not long after his graduation, he was dismissed from the army for drunkenness. For about five years, I was inspector of cadet uniforms and had to be at the cadet store nearly every day at 7:00 A.M. to supervise the fitting. I brought the overcoats lower and had them all about the same distance from the ground. I also had the cloth shrunk before being used. At one time, the white trousers disintegrated soon after being laundered. I made a long study and found that they were made from the sweepings of the mill. We discontinued the contract and went back to a former contractor, whose cloth had never given trouble.

My duties occupied me intensely during the day, and I generally went to the battery office at night. I had too little time for my family. I enjoyed teaching our little son to ride his bicycle, and I had him ride a horse with me as much as possible. We bought a light surrey and drove a battery horse, and on many afternoons we took short drives. On one occasion, the mare was not accustomed to such use and ran away. She was almost on a wagon in front and would no doubt have struck it, with disastrous results. My dear wife quickly pulled the rein and guided her away from the wagon. On another occasion, a gun team ran away and dashed into the trees in front of the campground. Although it looked very bad, I paid no attention, leaving the cadets who were the drivers and the cannoneers to take care of the situation. This pleased them very much, for they hated to be supervised and enjoyed overcoming difficulties alone.

My dear wife taught the very young children in the Sunday school conducted by cadets for the post children. When little Frances Christian[17] was asked who her teacher was, she replied: "I don't know, but Mrs. Summerall, Charles, and I are in the same class." It was a tribute to her selflessness and her understanding that made them feel a oneness with her.

Chapter 13

To Texas and Fort Myer

On March 11, 1911, I was promoted to major. A month later, I departed West Point to assume command of the Second Battalion, Third Field Artillery, whose station was at Fort Myer, Virginia, but the unit was then at San Antonio, Texas, as a part of the Maneuver Division.[1] It was with deep regret that I left the artillery detachment at West Point, then at its highest state of efficiency and morale. The noncommissioned officers were the equals of officers in the practical instruction of cadets. The artillery, both field and coast, commanded the respect of the cadets and was a popular choice of branch on graduation. The horses and equipment were equal or superior to those of the regular field artillery in the army.

The chief of field artillery in the War Department told me that he had assigned me to a "rotten battalion," which he expected me to change. When I joined the battalion, I found that his statement was conservative, for, although there were a few excellent officers, others were worse than useless. Some were very dissipated, including two captains. The men were uninstructed, the materiel in deplorable condition, and the horses much neglected. The Maneuver Division, composed of a large part of the army, had been assembled to invade Mexico as a result of border depredations. I at once began to establish camp sanitation, care of animals and equipment, and a course of instruction. The poor officers reacted with resistance or indifference.

Because the season was wet, the soft adobe made walking or movement very difficult. The troops were soon moved to Leon Springs for maneuvers. After my soldiers, batteries, and cadets at West Point, the contrast was distressing. Target practice was a farce because so many officers could not properly conduct fire. But the battalion improved rapidly as the marking and occupation of positions became very creditable. Yet the ignorance of even general officers of artillery, especially of indirect fire, shocked me. A brigadier general criticized me for placing the guns where the cannoneers could not see the target. I tried to explain to him defilade and indirect laying. His son, who had been a cadet under me at West Point, was the general's aide. When the general left, his son rode back to me and apologized for his father, saying that he knew nothing of indirect fire. On another occasion, a general told me that my guns were not pointing at the correct target. I asked him where he wanted them to point, and, on his giving the change in deflection, all guns were at once turned as he ordered. He appeared to be disgusted and rode away. The class from Fort Leavenworth had been sent to the Maneuver Division to assist in the latest training methods. Some of them told me that the colonels would not have anything to do with them. I realized the great truth that all impulses must come from the top. They cannot be passed upward. Although the troops suffered real hardship in the rain and mud, the experience was most useful.

The plan to invade Mexico was abandoned, and the battalion was ordered to proceed by rail to Fort Myer, Virginia. Each of the three batteries traveled by a special train and departed from San Antonio July 31, 1911. I traveled with Battery F, commanded by First Lieutenant Sherman Miles,[2] a most superior and loyal officer. Food was prepared and served in a kitchen car. I invited the young Pullman conductor to eat with the officers in the Pullman, but he felt too modest and declined. As we were entering Atlanta en route, he came to me and said quite simply that he had shot one of the porters. The porters were the most industrious and attentive whom I had ever seen. I asked him the trouble. He said that, when this particular porter did not obey at once when spoken to, he dragged the porter up by the collar when he resisted. Then he returned to the Pullman, got his revolver, and went back and shot the porter, who was attached to a standard sleeper. I went with him to where the incident occurred and found one porter dead and another badly wounded. The train conductor stopped the train and advised the Pullman conductor to escape. I advised him to remain and had the train conductor telephone the station to have the police and a doctor meet the train. The Pullman conductor was arraigned and acquitted the next morning. Unfortunately, he took our tickets with him, and we had some difficulty with his replacement.

On reaching Fort Myer, the men, horses, and guns occupied the best of buildings, in contrast to the exposure to rain, mud, and dust in Texas. I took a short leave and joined my wife and son at Markham, Virginia. I bought a pony for our son, and he began the happy experience of riding his own horse. My rank gave us a nice set of quarters near the flagstaff. I unpacked and settled the home and then brought my family from Markham. Here began one of the most difficult and trying periods of my service and one of the most valuable in accomplishment.

The headquarters and one squadron of the Fifteenth Cavalry occupied the post. The colonel[3] in command was the most worthless and contemptible officer I have known. His idea was to have the troops do as little as possible and let the officers run wild. This was incompatible with making a "rotten battalion" of artillery efficient. The post orders prohibited any duty after 12:00 noon, not even stables. The orders of the department commander were that troops would be trained as many hours a day as was necessary to make them efficient. I established an officer's school, which I conducted during the hours of morning stables, assigning one officer to supervise stables. Gunnery, materiel, battery, and battalion drills occupied the mornings. I devoted the afternoons to terrain exercises for the officers until stables. The post commander objected, and an inspector investigated complaints that I overworked the officers. I found that several of them spent the nights drinking in Washington, then were unfit for duty in the morning. Delinquent officers were called on to explain their offenses. Charges were preferred against one insubordinate lieutenant, but the post commander refused to try him. When I relieved him from my officers' class and he failed an examination for promotion, he complained that I had not instructed him. Another lieutenant cheated in a minor tactics problem that I gave to the post school class that I taught by copying an outpost map. He was tried and dismissed, after I preferred charges. I had tried to save this officer. After talking to him about his dissipation and neglect, he replied: "I would rather die than live the way you want me to live." The department commander sustained me against all complaints.

There was immediate and rapid improvement in discipline, instruction, gunnery, care of horses and equipment, and morale. The batteries took turns for the exhibition drills in the riding hall and reached a very high standard of performance. Each battery had a special team and harness for the funeral caisson,[4] which I inspected on each occasion. We had practice marches and overnight camps with night exercises in occupying positions and orientation. When the battalion took part in parades or exercises in Washington, the horses, harness, and carriages presented a beautiful appearance and attracted much favorable publicity and comment. The barracks were kept clean and neat, the messes were excellent, and the recreation rooms were attractive.

Our home was most comfortable, and my pay as a major enabled us to entertain as much as was reasonable. I bought a nice closed carriage and used two battery horses to pull it. The driver, a sergeant, took great pride in his position and would show off, especially at the White House receptions. My dear wife and I were on the post hop committee, and, though we knew little about drinks, when our turn came, we could provide champagne punch, which was drunk as fast as it was made. Our child went to school in Washington, taking an hour each way in the solid iron tire bus over the cobblestones. The children were tormented on the bus by a mean girl, and he had a violent teacher at school. Because of them, he would come home highly overwrought. Consequently, I often had his pony and my horse waiting his arrival so we would take a long ride, which he loved. One day, I acceded to his request and substituted a flat saddle for his McClellan saddle. The pony knew the change would enable him to get out of control. As soon as he left the post, he lowered his head, dragged the child out of the saddle, and bolted into the woods. One foot of the child caught in the stirrup, and he was about to be dragged to death when the stirrup strap slipped from the release. I caught the pony and give him a severe beating while the child cried and begged me not to punish the pony. I then mounted the pony, and we returned to the post. I rode the pony to the road and galloped him till he was worn out. He never gave me any more trouble. The child loved him dearly.

At the time, the army had a requirement for all officers to take a ninety-mile ride in three days, of six hours a day.[5] General Frederick D. Grant[6] and his staff came to the post for their ride, and we gave them a buffet supper, which they enjoyed. Our little son became a Tenderfoot Boy Scout and a bugler. He made a lovely picture in his uniform with his bugle by his side. A nice boy came in the evenings at weekends and taught him scouting. A lovely young girl, Gilberta Hawkins, the daughter of Mr. and Mrs. Hawkins, whom we met on the ship from Skagway, spent Christmas with us each year from her boarding school at Ogontz, Pennsylvania. She and our child loved each other. Our Christmas decorations and tree were always gay, and the time was a happy one.

I went to the School of Fire for Field Artillery at Fort Sill, Oklahoma, for a three-month course during the winter of 1911–1912. When I returned to Fort Myer, the post commander[7] wrote to the commandant at the school and asked if I had been proficient. He was looking for anything to hurt me. While he hated me, I always treated him with respect, which I did not feel. One day, he sent for me and said that the department orders required him to have a tactical ride of all officers at the post once a week. He said that I should conduct the ride, leaving post headquarters at 12:00, and reporting back with the officers at 12:15. Of course it was a farce, but I obeyed his order.

The Militia Bureau of the War Department wanted me to establish a summer camp and train the national guard field artillery east of the Mississippi River. I searched for sites on the Cumberland Plateau near Tullahoma, Tennessee, in the Valley of Virginia, and at the Pink Beds in the Pisgah Mountains of North Carolina, but nothing was suitable. Finally, General Leonard Wood, the chief of staff, suggested that I investigate the Pocono Mountains of Pennsylvania. There, I found a vast plateau, fourteen hundred feet high, covered with the most forbidding rock, gullies, extensive bogs, jungle, and impenetrable forest of second-growth hardwood trees. The great pine forest had been cut off about 1865. The area was about thirty miles in extent and mostly owned by the Lehigh Coal and Navigation Company. There were few roads and no settlements in the area. I contacted the company and obtained permission to fire over the area in the summer of 1912. The only possible campsite was about three miles from the village of Tobyhanna at the eastern edge of the area. The War Department furnished a small allotment of money, with which I arranged with the company to establish a camp.

Early in the summer of 1912, my battalion marched to near New Haven, Connecticut, for extensive maneuvers with the troops in the Eastern Department. At the end of the maneuvers, the battalion went by train to Tobyhanna. We arrived at night, detrained, and marched three miles to the desolate campsite. There was only water enough in a small spring to make coffee for breakfast. The horses were taken first to a muddy lake and then to the creek at Tobyhanna for water. Officers and men were overwhelmed by such hardship. I procured barrels of water for use at the camp and immediately began arrangements for target practice. The rocks and the jungle awed everyone, but the men soon learned to deal with them. We located target sites and firing positions and had a most instructive experience in reconnaissance, location of batteries, determining firing data, and the conduct of fire. When we finished, it was a different command, and it felt that it could overcome any difficulties in the field and fire effectively. The nights were always cold and the days cool, but the health of the command was excellent. We returned by train to Fort Myer. My enemy the post commander reported to the War Department that I had seriously broken and damaged the battery carriages. Because this was false, I heard nothing of it.

The Militia Bureau gave me a little money to rent a campsite on the edge of Tobyhanna and to bore a well and build a water tank. During the winter, I intensified the training and much smoke bomb practice and occupation of positions in the country. At the inauguration parade for President Woodrow Wilson on March 4, 1913, the superb appearance of the battalion "swept everything before it."

Early in the summer of 1913, the battalion marched for Tobyhanna. The post commander detained three of the officers on special duty. At my first camp en route, I telegraphed department headquarters to have them join their batteries, which they did. We had no personal contacts, yet I made them perform their duties or explain why. During the march, the men of my old battery straggled and trespassed on private property. I made the worthless captain read the Articles of War to them and confined the battery's officers and men to camp for the rest of the march. After that, I had no more disciplinary troubles in the battalion. The men and especially my headquarters detachment were anxious to do all that I required.

We soon cleared the rocks and jungle from the campsite, built roads, operated a rock crusher, and built sheds for kitchens and stables. A succession of national guard batteries from Massachusetts to Georgia occupied a part of the camp for a week each. The New York field artillery came by regiments. They were shocked at first by the rocks and jungle, but they soon learned to overcome any obstacles. The officers and men were superior in intelligence and morale. They received intensive instruction in caring for horses, harness, and carriages for service of the guns, maneuvering and occupying positions, map problems, smoke bomb problems, and target practice with service ammunition. In the evenings, we held a critique, in the large mess hall, of the day's work. They were eager to learn and profited by all that was said.

When a young captain from Georgia was told that his men must groom and care for the horses, he objected and said that that was work for Negroes. I convinced him to the contrary. At that time, none of the national guard batteries had any horses. This work for my battalion was very stimulating and added to its morale and efficiency. The discipline of the soldiers was excellent. When we finished with the national guard, we had our own target practice, which was excellent. I believe that it became the most efficient battalion in the army and was highly rated by the inspectors. My little family and the wives of other officers found comfortable boarding-houses at Tobyhanna and spent the summer near us. We returned in October by marching to Fort Myer. En route, we stopped in Columbia, Pennsylvania, for Old Home Week, where the people entertained the officers and men and we made many warm friends.

Again, the garrison program was devoted to thorough training, with special emphasis on qualifying gunners. The usual weekly exhibition drills in the riding hall kept horses and vehicles in show condition. The presence of soldiers had greatly disturbed the Quaker residents of Pocono Lake, which was a summer resort for people in Philadelphia, located about five miles from Tobyhanna. Mr. A. Mitchell Palmer,[8] who was a Quaker and the attorney for the Lehigh Coal and Navigation Company, told me that, unless I could placate them, we might not be allowed to use the range. General Thomas H. Barry, who had been superintendent at West Point when I was on duty there, commanded the Eastern Department. On my application, he ordered the Fifteenth Cavalry band from Fort Myer to the camp at Tobyhanna during the summer of 1914. The band concerts, especially on Sunday afternoons, attracted hundreds of visitors, including the Quakers. Among them were the Quaker mayor of Philadelphia and his wife, who became very friendly to me. They invited me to their home in Pocono Lake for noon dinner one Sunday, and the Quaker population returning from church saw me sitting on their porch. After that, I had no further opposition.

In early summer 1914, the battalion marched to Tobyhanna and prepared the camp. By that time, a railroad siding had been extended to the camp, a large warehouse constructed, more roads built, and the camp area improved. During the summer, national guard field artillery came by regiments, battalions, or batteries from Massachusetts, Rhode Island, Connecticut, New York, Pennsylvania, Ohio, Virginia, the District of Columbia, Georgia, and Louisiana. To urge them on, I told them that the guns of the next war were being aimed at Tobyhanna. I also told them that, if they could maneuver guns at Tobyhanna, they could do so anywhere in the world. In a few years, they learned that both statements were correct. None of these troops were ever baffled by the difficulties in France in World War I. We duplicated as nearly as possible the School of Fire at Fort Sill. Many officers on detached service came to the camp for a short course in gunnery and target practice. A group of marine officers attended and learned enough to incorporate field artillery in the marines.

Initially, we had no thought of war during the summer of 1914. The Carnegie Foundation[9] had published a statement that it had abolished war and must find some other use for its money. A portable field radio of a new design had been issued to the camp for experimental use. On Saturday nights, the officers had dances in the large mess hall. During a dance one night, the radio operator rushed to me with messages that he had picked up, ordering all British ships to the nearest American ports, and announcing the recall of ambassadors of the European antagonists to their homes. We knew that war was a reality. The First New York Field Artillery Regiment, commanded by Colonel H. H. Rogers Jr.,[10] was in camp. Many of the officers were brokers and left at once to meet the crisis in the stock exchange. After the national guard finished, General Barry and his staff[11] came to the camp to take the ninety-mile ride with the officers of the battalion. On our marches, we usually camped at South Bethlehem, Pennsylvania, where we made friends with the executives of the steel company. During the summer, the company president, Mr. Eugene Grace,[12] and his executives visited the camp for a day.

When we were at Tobyhanna, my many duties required that I be constantly going on the range and in various places about the camp. Consequently, the men in my battalion thought that I was ubiquitous. One day, a visitor asked one of my officers where I was. He replied: "Do something wrong in the camp, and he will be right there." An officer told me that General Barry, while talking to a group of officers about me, said: "Summerall is the best G——D—— officer in the army."

The veterinary surgeon, Captain Griffin,[13] was musical, and in the evenings he sang service songs, including "The Caissons Go Rolling Along." He contributed much to morale. He also wrote some Tobyhanna songs, which all learned and enjoyed. Many visitors came from hotels in the neighboring summer resorts to witness the firing and to lunch. Among them was Miss Margaret Wilson, the president's daughter. In turn, they invited the officers of the battalion to dinner and dances on the weekends. Liberal passes were given to the men, who visited the towns and made friends. The work was hard, but there was sufficient recreation to maintain high morale. The greatest contributing factor was pride in the efficiency and superior training and standards of the battalion. It was no longer "rotten." It was ready for action anywhere, anytime, and under any conditions.

Again, our families boarded near our camp, so we could see much of them. It was my greatest happiness to be with my dear wife and child in the evenings or to have them come to camp. When we broke camp October 18, 1914, to march back to Fort Myer, there had been days of sleet, snow, and ice. However, on making camp at Stroudsburg down the mountain, the temperature was mild.

Chapter 14

The War Department

On September 4, 1914, I reported to the War Department as assistant to the chief of the Militia Bureau in charge of the field artillery of the national guard. The War Department budget was being completed, but I managed to have it include $600,000 for horses for the national guard field artillery. The idea was ridiculed, but I believed that I could make a good defense before the Committee on Military Affairs of the House of Representatives.[1] One of my responsibilities was to enforce proper care of and accounting for the materiel used by the militia units in most states. There were many callers from all over the country with difficult problems. The abilities of the inspector instructors[2] varied, but generally they were good officers. They usually reported that the militia batteries were untrained and in poor condition. Correspondence was very heavy, and I often went to the office at night. Soon, I moved my family to an apartment in the Westmoreland,[3] where we were very comfortable. My dear wife's father and stepmother occupied an apartment in the same building.

The war in Europe caused much interest, and requests poured in for the organization of new militia batteries, which required additional firing areas. I began a search for camps and target ranges throughout the country. The reservation at Sparta, Wisconsin, was satisfactory for the batteries in the north-central states. I visited Anniston, Alabama, where I found about eighteen thousand acres of poor and wasteland that was admirably suited for batteries in the south. I knew of a very good tract of at least sixteen thousand acres near Monterey, California, where the troops had maneuvered when I was stationed at the Presidio of San Francisco. For the summer of 1915, I arranged an intensive program for the artillery in the east and south. When I was called before the Military Affairs Committee to defend the money for horses, I had thought out what to say. The chairman of the committee was noted for his objection to such appropriations. Before I had proceeded far, he stopped me and said:

"That sounds good to me." As soon as the appropriation became a law, I arranged for stables, the purchase of horses, and the detailing of caretakers from the army. It was a new day for the national guard artillery, and the response was gratifying.

In 1915, I became a member of the Board of Ordnance and Fortification, which often took me to Sandy Hook Proving Ground. I also conducted terrain exercises for the class at the Army War College. In the budget for 1916–1917, I had an item of $600,000 to purchase national guard ranges and camps at Tobyhanna, Anniston, and Monterey. This was appropriated, and I proceeded to negotiate terms. For the amount appropriated, I procured about twenty-eight thousand acres at Tobyhanna, including a large area for a campsite, sixteen thousand acres at Anniston, with about two thousand acres of government land adjoining, and sixteen thousand acres of a Spanish grant on Monterey Bay near the Del Monte Hotel.

Early in 1917, I went to California, encountering a blizzard en route. While in San Francisco, I became very ill and remained in Letterman Hospital for three or four weeks. When subsequently I offered a price for the Jacques property at Monterey, the agent for the owners declined. I then went to see the lawyer for the owners of the property, who were two young boys and a young sister. He occupied a shabby office near the Barbary Coast and himself looked very scrawny and shabby. When I introduced myself, he told me to leave. I demurred and asked him where he came from. He said Virginia. I told him that I knew Virginia and had marched through it. He exclaimed: "Do you know about Virginia ham?" I said yes. He opened a bottom drawer in his desk and took out a ham bone with very little meat remaining. He said: "What is this?" I said: "That is a Virginia ham." He said: "I read a recipe about cooking a ham. It said 'Take a ham.' It should have said 'Take a Virginia ham.'" I then asked him to lunch with me and took him to the Palace Hotel. I told him to order whatever he wanted to eat and drink. He did both. When we walked out he said: "At last, I have had something at the expense of the government." When we returned to his office, I offered him $160,000 for the sixteen thousand acres. Because he had the power of attorney for the owners, he signed a purchase agreement for the property, which I wrote.

I returned from California by way of Alabama, where the Alabama National Guard artillery and a regular battery used the Anniston range that summer. On reaching there, I learned that a board of officers would inspect the site for a mobilization training camp for the war. The senior officer of the board told me that they would reject the site, which they had not seen. After I went with them to see the land, they recommended it. It was a large training camp during World War I and became a permanent post. The Monterey camp became Fort Fremont in World War I and Fort Ord in World War II. Tobyhanna was used for tank training in World War I but appears not to have been used in World War II.[4]

In 1916, I was made recorder of the congressional commission to investigate the resources of the country for manufacturing arms and munitions, and, in 1917, I was made a member of the Board of Ordnance and Fortification. As a member of the Ordnance Board, I tested the new 3.6-inch howitzer and recommended its adoption. It corresponded to the French 105-millimeter howitzer, which was not adopted until World War II, although I fought for it during and after World War I. A number of batteries were added to the national guard, and 1917 found most of them in a reasonable to excellent state of training. I had a wide and friendly acquaintance with the officers of these units, and I visited as many of the organizations as possible. Among the requests for new batteries was one by the adjutant general of Connecticut for a battery at Yale University. But Yale would not supply an armory and stables, which were later donated by an alumnus. I sent Major R. M. Danford[5] (afterward major general) as instructor there. He turned the Yale students into an excellent battery. When the Mexican border troubles began in 1916, I expanded the Yale battery to a regiment and ordered it to Tobyhanna for training. It became a training school for officers in 1917. When the Pennsylvania National Guard field artillery was ordered to the Mexican border, they wanted me to be a brigadier general and command the brigade. I declined as this belonged to one of their colonels.

With the Munitions Board, I visited most of the factories in the country that could produce war material. Morgan and Company allowed me to study their experience in contracting for and delivering guns and ammunition to the Allies. I reported the time required for manufacture and for full production of every class, which our wartime experience verified.

In 1916, I was a member of a board to recommend a reorganization of the field artillery. It was called the Treat Board after the president of the board, Colonel C. G. Treat.[6] I made extensive studies of the experience in the European war and prepared a brigade organization for each division consisting of two regiments of 3-inch guns and one regiment of 3.6-inch howitzers. Brigades of larger caliber howitzers and guns were to be organized as corps and army artillery with divisional brigades in corps and army reserve. It was adopted except that 155-millimeter howitzers were substituted for the 105-millimeter or 3.6-inch howitzers. It was the most efficient organization of artillery in the war in any army and did much to win our battles. For my part in doing away with an inefficient organization previously adopted, I made an enemy of a member of General Pershing's staff, who tried to injure me during the war.

One day, the chief of staff sent for me and stated in much perturbation that the chief of ordnance had informed him that it would take nine years to produce the 3.6-inch howitzer that I had tested. I told him that it could be produced in not over eighteen months and perhaps in a year. This was the first of a succession of conflicts with the Ordnance Department that I have never been able to understand.

While everyone expected us to enter the war, there were many who felt that it was not our war and that we would suffer for people who never had been and never would be our friends. Many others were radically pro-British, calling everyone a traitor who dared think otherwise. The people we saw at the apartment house were loud in their partisanship, and some officers dared to oppose our entering the war. We tried to avoid taking either side. My duty was to prepare for war and to respond to any order. When the *Lusitania* was sunk, we knew that war was inevitable. In preparation for mobilization, I was appointed as a member of the board to locate and recommend training and mobilization camps.

Chapter 15

The World War

Toward the end of June 1917, the War Department informed me that I would go to England and France as a member of a commission to study the types and the employment of field artillery.[1] The chief of staff[2] sent for me and in great secrecy handed me a letter addressed to "General John J. Pershing. For his eyes alone." He told me to be in Halifax, Nova Scotia, the next morning and give it to General Pershing. I was ordered not to let anyone know that I was going and to sail from Halifax for England. In the hall, I met a member of the board on camps, which was about to convene, but I did not tell him that I would be absent. Without reporting my departure to the chief of the Militia Bureau,[3] I packed my trunk and left my office. I had told my dear wife that I was going to inspect a camp. When she went with me to the station, I felt that she knew.

I reached Halifax early the next morning and went at once to seek information at the American consulate. Although I knew nothing of General Pershing's movements, others understood that he had sailed from New York on the SS *Baltic*.[4] When I asked for the consul, I told him who I was and that I must meet General Pershing to give him a letter. He told me that he knew nothing of General Pershing and withdrew. He soon returned and seemed more friendly. I assumed that he had looked me up in the *Army Register*. He then said that he was translating a code message about General Pershing but would go with me to see the senior British naval officer in port, who was a captain of one of the station cruisers. The captain received me coldly, insisted that he knew nothing about General Pershing, and declared that, if he did know, he would not tell me.

By that time, the consul had become very friendly. I had avoided registering at a hotel and had seen no one except the consul and the British captain. The consul took me to the hotel for lunch. One of the first people I saw was Captain Quekemeyer,[5] General Pershing's aide. He had been a cadet while I was on duty at West Point and was very friendly. He said that he had been looking for me. I asked how he knew that I was in Halifax. He replied that everyone in Halifax knew that I was there. He said that he had come to join General Pershing when he arrived. On entering the dining room, the consul introduced me to Captain Hayes,[6] who commanded the *Olympic*,[7] lying in the stream loaded with Canadian troops ready to sail for Europe. Meeting him was very fortunate for me later. I spent the remainder of the day with the consul at his office and went home with him that night. He constantly called the port but could get no news of the *Baltic*. About midnight, the telephone rang, and the admiral's chief of staff told the consul that the admiral had just arrived and was informed that the consul wanted to see him. The consul explained my mission. The chief of staff said that the *Baltic* had not reported and was not expected to call. I told him that I must see General Pershing and asked him to send me in a destroyer to locate the *Baltic*. He said that this could not be done. I was chagrined and bitterly disappointed at the failure of my mission.

The next morning, the consul and I went to the port office. The captain whom we had seen at lunch the previous day was very friendly, explaining that, because enemy submarines had been sighted off Halifax, the *Baltic* had not reported or answered my radio and had been ordered to proceed from wherever she was without calling at Halifax. I learned that Mr. Balfour's[8] party would sail that day on the *Olympic* and asked for transportation on her. The captain of the port radioed the British military attaché in Washington for authority for me to proceed on the *Olympic*. During the afternoon, I was told that I could board the *Olympic*, and I did so at once.

On the ship, I met the members of my military mission, who had joined Mr. Balfour in Ottawa and accompanied him to Halifax. I occupied a cabin with Captain Quekemeyer and kept the letter to General Pershing sewed in my vest pocket, wearing the vest when I slept. The voyage was uneventful.[9] Destroyers escorted us two days out from Halifax and met us two days out from Liverpool. The ship maintained a speed of about thirty knots and continually zigzagged. Boat drills were held, but there was no uneasiness. We saw little of the Canadian officers or of Mr. Balfour's party. We could not fail to see that they looked down on the Americans. Our mission consisted of three American field artillery officers and one marine artillery officer, Major Dunlap,[10] who was with me in China and who had trained at my camp at Tobyhanna.

We reached London twenty-four hours after General Pershing,

and I first saw him on entering the lobby of the hotel. I knew him very slightly. We had been ushers at a wedding when he was a first lieutenant and I was a second lieutenant, and he and his wife had called on us at Fort Myer when I was a captain. I told him I had a letter for him, which I had tried to deliver in Halifax. He took me to his room, where I ripped the letter from my vest pocket and handed it to him. He read it, waited a moment, and made a grunt. I saw that the interview was finished and left. I never learned the contents, but I have surmised that his going on the *Baltic* was made public so that he could transfer to the fast *Olympic* in Halifax and, thus, deceive the German submarines. Fortunately, the slow *Baltic* made the trip successfully.

The American military attaché in London, Colonel Lassiter,[11] had prepared a schedule for us to visit Aldershot and the principal training camps, especially for field artillery. We also found a number of invitations to dinners, including one from Lady Astor[12] and a very formal one from Lord Curzon[13] to a dinner at St. James's. I felt that our mission was essentially professional and not social. I could not accept Lord Curzon's dinner without sacrificing a part of the itinerary, so I regretted. I could be in London for Lady Astor's dinner, and we all accepted. Some of the party accepted Lord Curzon's dinner, including a clerk in the War Department who had been commissioned a major to do our clerical work but who promptly assumed the status of his rank without credit to it.

We were cordially treated at the different camps and spent one night at the home of the commanding general of the southern part of England in Salisbury. We learned much as to types of artillery, training, methods of employment, and the auxiliary services. Up to that time, the British and the French had refused to give us any information. I made copious notes, much as I had recently done as a recorder of the Treat Board for reorganizing our field artillery and as recorder of the Board on Munitions authorized by Congress. I was gratified to find many of my views vindicated in the war, especially as to the necessity for the 105-millimeter howitzer to replace the 75-millimeter gun. Signal communications and the location of hostile batteries by triangulation of the record of hostile shots were also most helpful.

The dinner at Lady Astor's was a most remarkable and happy occasion. The guests included the officers of our artillery mission and several American ladies who had married into the British nobility. Lady Astor was charming, natural, and cordial. The other ladies appeared to become once more American, and we talked as though we were a family reunion. After dinner, we sat informally in a small reception room and sang American folk songs like "Ol' Virginny," "Ol' Kentucky Home," "S'wanee River," "Casey Jones," and others. Lady Astor's sister, at the piano, led us, and all found voice and, I believe, emotion. I felt that love of native land can never die and longs to express itself with kindred souls.

While in London, I met some officers whom I had known very pleasantly in China. We called at the War Office and were cordially received by Lord Derby,[14] the minister of war. We also attended a reception at the American embassy, where Ambassador Page made us most welcome. There I met Lord Roberts,[15] England's first soldier, and other prominent men.

The hotels were gay with many British officers in their handsome blue uniforms and elaborately dressed women. No doubt, everything was being done to give officers and men on leave a maximum of pleasure and to keep up morale. I recalled that, in the dark days of the Confederate army, General Lee urged the people of Richmond to entertain the soldiers and be as gay as possible. We knew that these were dark days for England and that, but for the entry of the American army, the war would be disastrous. Our simple and inexpensive uniforms compared unfavorably with the British. General Pershing at once adopted the Sam Browne[16] belt, and his staff were soon wearing London-made uniforms.

In order to comply with the orders of the chief of staff, I could not write to my dear wife until I reached Liverpool. My letters were censored, but I told her of my activities as frequently as possible. I realized that my going caused her great anxiety, but her courage and resignation were sublime and heartened me.

When we crossed the Channel to France, we at once visited the British front at Messines. This had been a battle of great magnitude in which the British had just captured several miles of territory. The artillery action was continuing with much intensity while the troops consolidated their gains. The British showed us the preliminary plans and the reduced terrain where the attack had been rehearsed. We were especially interested in the sound-ranging system for locating enemy batteries and saw it in action. The liaison system between artillery and infantry also was of interest. We were impressed by the desire of the British to have the 105-millimeter howitzer substituted for their field guns, which were not powerful enough. The chief of staff of the Fifth British Army[17] was especially cordial and took pains to explain the operations of army headquarters.

At first, we dreaded becoming infected with lice, which were said to infest all billets, and we had taken some powder from London to use as a preventive. The powder burned me severely, and I had to risk the lice, which we escaped.

The last night of our stay with the British, they prepared a practical joke on one member of our party. He wore the U.S. Marine Corps uniform, which attracted much attention by reason of its resemblance to the German uniform. Accordingly, the British military police arrested him when we were in a party after dinner. He turned to me for identification, and I expressed doubt. His reproachful look was distressing. At last, the police said that the escaped German prisoner for whom they were looking had a scar between his shoulders, whereupon the officer removed his shirt to prove his innocence. When he saw the joke, he enjoyed it with us. It relieved the tension under which we were laboring.

We then proceeded to Paris. The American military attaché, Colonel Margetts,[18] had been one of my lieutenants in Skagway, Alaska. He had made reservations for us at the Hotel Crillon. French liaison officers, under Major Reille,[19] at once took charge of us, and we proceeded to the artillery training school at Vincennes, where we were impressed by the fine horses as well as the heavy 155-millimeter guns and howitzers. Miniature trench systems and terrain were used to illustrate deployments, defense, and attack methods. We also saw the palace with its trophies and living rooms of Napoléon and his empress. The beautiful and expansive hunting park was the reverse of war.

We were then taken to a number of places along the French front, and it was evident that the program provided for as many of the French troops as possible to see us so as to show them that America was in the war. We witnessed barrage fire as well as the camouflaging of battery positions and roads. We were impressed by the readiness of batteries to lay down instantaneous protective fire for the infantry in the advance trenches on their sending up signal rockets. While the French were proud of the 75-millimeter gun, they all wanted the 105-millimeter howitzer to add power with the same mobility. They wanted the 155-millimeter guns and howitzers for their especial power, but they realized that they did not have the mobility for divisional artillery. All this, as well as the British view, confirmed my decision on the Treat Board to have one regiment of 105-millimeter howitzers and two regiments of light guns in the divisional artillery brigade.

We lunched with General Gouraud,[20] who commanded the Fifth French Army with headquarters at Chalons-sur-Marne. He was the most impressive figure whom we visited. At Gallipoli, he had lost an arm and was seriously crippled in one leg by wounds. His face was much like the prevailing pictures of the Christ. Although he spoke no English, he chose words much alike in both languages and spoke slowly so that we understood him. He was very close to his men. We were told that, after a battle, he would have one soldier from each company at a dinner. He would proceed from table to table and shake hands with and speak to each man in friendly and complimentary terms.

At the table were a number of French generals and officers. I sat on his right. I asked him if the officers there were members of his staff. He said that they were commanders of different units and a few officers to be decorated after lunch. He said that he had some of his officers to lunch every day and added: "The best liaison is the dinner table." In this short period we were together, I learned much about leadership and the human element in war from him. Everywhere, the French served excellent wines, which I could not appreciate and did not drink. No doubt, they considered me what they came to call us, one of the *sauvage Americaines.*

We returned to Paris on July 2 and found the city gaily decorated with flags and bunting for July 4. On July 3, we were invited to a conference with General Pershing's staff, although they had no control over us.[21] All of the staff and General Pershing were present at his headquarters in the Mills home.[22] I was asked to state the result of our visits to the British and French armies. I thus pointed out the superior organization of our artillery brigade and the general advocacy of the 105-millimeter howitzer and recommended a chief of artillery for corps and armies, which both the French and the British had. I urged the Allied estimate for the attack of one 75-millimeter gun for every fifteen yards of front, one 155-millimeter howitzer for every fifty yards of front, and one 155-millimeter rifle for every hundred yards of front. I was at once viciously attacked, personally and officially, by officers of the staff, whom I hardly knew, for trying to promote a lot of artillery generals, for advocating the light howitzer, and for such a quantity of artillery. I replied with equal force and resentment. I told them that the infantry would pay in losses for lack of artillery. That is what happened. General Pershing then asked me to step out on the porch and said: "Summerall, I want you to get together with my staff." I replied: "General, no one wants your success more than I. Your staff was wrong, and I am going to Washington and fight for what I know is best for our artillery." He seemed to be furious and returned to the house without a word. I was then told by the staff that they wanted a report from me on July 5. I procured a stenographer and, while all of Paris was wildly celebrating the day, worked all of the fourth on the report. On the fifth, I was told there would be no conference, and I left a copy of my preliminary report with the staff. I felt that I would not be allowed to return to France and that my part in the war was ended.

We gave a farewell dinner at the Hotel Crillon to the French liaison officers and left for England to return home. Before leaving Paris, I bought a five-pound box of loaf sugar at the American commissary, which is all they would let me have. I knew that there was no sugar in England, and I wanted to take this to Lady Astor as a small token of appreciation. On the train to the port, we locked ourselves in a compartment to exclude strangers. There were several knocks at the door, to which we did not respond. Then the knocks were followed by a stream of profanity and abuse, and we knew it was an American who would not be denied. We let him in and soon became good friends. He was an officer of the air service. At Cherbourg, the British transportation officer was most discourteous and gave us much trouble about our passage on the Channel boat. I had some hot words with him, and we were finally allowed to go on board. There was no doubt in my mind that the British had only contempt for Americans.

As soon as we reached London, I called at Lady Astor's with the sugar. The footman said the she was at the hospital, so I left the box with my card, knowing that she would give it to her wounded. We sailed from Liverpool on the USS *St. Louis.* After proceeding down the Channel, the ship turned around after dark and steamed around the north of Ireland, in order, no doubt, to deceive the enemy submarines. On board, I was able to have one of Lord Northcliff's[23] stenographers, who was coming to join his propaganda staff in the United States, to take my notes and write my report, which I submitted on reaching Washington. Of course, the reunion with my dear wife and child was the great event of my return.

Probably the closest friend to Mr. Newton D. Baker, the secretary of war, was Mr. Benedict Crowell,[24] with whom I had been associated as a member of the Munitions Board, and for whom I felt a strong liking and respect. On returning to Washington, I at once asked him to lunch with me and explained the nature of my report and the unfortunate experience with General Headquarters (GHQ) in Paris. I asked him to acquaint the secretary of war with the situation. He said that he would have the secretary dine on his yacht that night and discuss the matter with him. The next day, the chief of staff[25] sent for me and stated that the secretary of war had directed him to appoint me as president of a board on types of artillery for the war and to have me select the other members. I did so, and the board recommended that the 105-millimeter howitzer, already developed by the Ordnance Department, be put into immediate production for divisional artillery. I also prepared a program of instruction for the School of Fire for Field Artillery and for artillery training centers.

Chapter 16

Over There

Early in September, the War Department appointed me a brigadier general and ordered me to take command of the Sixty-seventh Field Artillery Brigade of the Forty-second Division at Camp Mills, Long Island. I assumed the command on September 5. The brigade consisted of the 149th Field Artillery, National Guard of Illinois, Colonel Henry J. Reilly[1] commanding; the 150th Field Artillery, National Guard of Indiana, Colonel Robert H. Tyndall[2] commanding; and the 151st Field Artillery, National Guard of Minnesota, Colonel George E. Leach[3] commanding. Officers and men were superior in character and ability, with many young college men in some of the batteries. The highest morale and pride prevailed. But only the personnel were at Camp Mills, as horses and materiel were to be furnished by the French on arrival overseas. Thus, the time before sailing was devoted to theoretical training, dismounted drills and marches, camp sanitation, and guard duty.

During these days of preparation, my dear wife and son remained nearby at the hotel in Garden City where I could see them every evening. Among friends who stayed there was Mrs. Arthur MacArthur, the mother of Colonel Douglas MacArthur, chief of staff of the Forty-second Division. She made particular efforts to be friendly with my wife. Mrs. Henry J. Reilly, the widow of Captain Reilly, who was killed in Peking, and the mother of Colonel Reilly of the 149th Field Artillery, was also there, and we saw much of her. A number of the families living near the camp were most hospitable and entertained the officers and men at their beautiful homes. Orders soon came for the brigade to proceed to France aboard the SS *President Lincoln*. We boarded her at night and sailed October 18, 1917. It was at once evident that the ship was not prepared for such a trip. The brigade numbered about five thousand officers and men, while the ships' company added fifteen hundred more. The ship, an eighteen-thousand-ton German freighter built in 1912, had only a dozen small lifeboats. She had been reconditioned hurriedly, with piles of crates, containing empty five-gallon cans, prepared as life rafts. The men were placed in all five decks to the bottom of the ship and were so crowded that there were standees waiting for beds. There were compartments on each deck with doors that could be made watertight.

I immediately had a conference with Captain Yates Stirling,[4] his executive, Commander Foote,[5] and the navigator, Lieutenant Isaacs,[6] of the navy. In the event of a torpedo attack, we prepared a program of alarms and detachments to lower the rafts and tie the lines to the railing on the side of the ship. The men were to climb down rope ladders and grasp the crates of cans. Sentinels, placed at the doors of all compartments, would close them when the men had left the compartments at the alarm whistles. Orders to abandon ship would be given to me by the captain on the bridge, and I would transmit them by aides to the regimental colonels. The batteries were to form and remain at attention at the sounding of the alarm whistles. We assumed that any enemy submarine would follow the ship's zigzag during the night, at dawn move ahead of the ship, and fire torpedoes as the ship passed the submarine. We guessed that the ship might be struck by two torpedoes, would careen on the side hit, but would float for one hour. That would give time for the men to lower the rafts on the lower side, climb down the rope ladders, and gain a hold on the crates, which would at once be lowered and tied together, with the last one tied to the rail of the ship. Men were ordered not to jump and to wear life preservers at all times. To prepare for an emergency, the crew sounded the alarms every day at various hours, and eventually the men could form on deck in less than five minutes. A battalion of colored grave registration troops, who could not be trusted on deck, formed between decks with orders to go on deck only when instructed. Subsequent to our voyage, the SS *President Lincoln* was torpedoed with fifteen hundred men on board. All of the phenomena predicted occurred except that she was hit by four torpedoes and floated forty minutes. Only eighteen men were lost, probably the sentinels over the compartment doors. Captain Foote, the former executive officer, was then in command, while the navigator, Lieutenant Isaacs, was taken a prisoner by the German submarine.[7]

Except for seasickness, which resulted in vomiting everywhere, our voyage was uneventful. I inspected the ship daily. On the first day, I managed to see one section of the five decks. My staff and the ship's officers followed, making a long column that was still starting below when I ascended to the top deck. As I descended the first flight of steps to begin another section, I realized that I was about to vomit. When I started back as I had come, the sentinel very properly obeyed his orders and told me to ascend by the adjoining steps to preserve one-way traffic. I barely reached the rail when all inside of me came up. I was deathly sick. At last I managed to turn around and behold the staff of about fifteen officers standing at attention behind me. I managed to say: "That will be all today, gentlemen." They saluted, and I struggled to my berth. Lieutenant A. B. Butler,[8] my aide, at once recorded it in a caricature or cartoon and issued the scene in the day's mimeograph for the amusement of all.

The day before we reached St. Nazaire, the engine failed owing to sabotage by the Germans before the ship was seized. We dropped out of the convoy and saw it disappear with the escort cruiser and destroyers. We might easily have been torpedoed, but timely repairs enabled us to proceed alone, and we reached St. Nazaire to our great relief on October 31.[9] We unloaded the troops and went into camp. At the last dinner on board, my officers asked me to speak to them. I tried to lay down a few rules of conduct and leadership, but my main injunction was for the officers to be themselves whatever they wanted their men to become. I believe that in the campaign that followed, they heeded my words. After a few days, we proceeded by marching and rail to Camp Coetquidan, an artillery training camp, in Brittany.

From the time we landed, we were impressed by the unfriendly attitude of the French. It was evident that they wanted the war to end and resented our coming to prolong it. There is no doubt that the French were completely defeated mentally and physically and would have surrendered at any moment but for the arrival of the American army.[10]

At Camp Coetquidan, the French issued horses, equipment, and guns to us.[11] The 149th and 151st Regiments received 75-millimeter guns, and the 150th Regiment received 155-millimeter howitzers, which had been adopted by GHQ for the divisional artillery. At St. Nazaire, I received a telegram indicating that the secretary of war had revoked his approval of my board's recommendation to put the 105-millimeter howitzer into production. Although the 155 (which the secretary favored) had great power, it lacked mobility for a division.

Soon an intensive schedule of training was begun under French officers and American officers who had attended the French artillery schools. Progress was rapid, and target practice soon showed much proficiency.

A large number of German prisoners were at Coetquidan. When they moved to new barracks, we were placed in their old barracks, which were infested with lice and vermin. While talking to an officer who was a prominent and wealthy lawyer in Chicago and who later became an ambassador,[12] I noticed lice crawling around his collar. He, no doubt, saw them on me. At once, a malignant form of measles broke out, with most cases turning into fatal pneumonia. The young college boys especially became victims. Because there were no hospital and no nurses, the sick were placed in an old barrack or in sheds without heat or sanitation unfit for occupancy. Many died. Every afternoon at four o'clock, the caissons would form at the morgue and take these young bodies to a cemetery, which we started against the protests of the French commanding officer. The surgeon objected to having nurses in camp and refused to ask for them. The enlisted men were ignorant and wholly incompetent to care for the sick. I went to the corps headquarters, where I knew the chief surgeon and explained the situation. After a Pittsburgh unit of nurses,[13] which had just arrived, was sent to the camp, conditions promptly improved.

The French commander of the region with headquarters at Rennes, General D'Amade,[14] called, and we became good friends. He had distinguished himself at Gallipoli. I had just received some cigars, which he greatly appreciated. The mayor of Rennes invited me and one officer and one soldier of every rank to a reception at the Hotel de Ville.[15] I made an address from a balcony to a multitude of people. A professor from the university translated my English into French. There was much champagne and abundant refreshments. We were shown the noted tapestries depicting historical events of Brittany. I inspected the new troops in training, consisting of very young and rather weak boys. It was evident why France could no longer fight. I remarked to General D'Amade that these boys were not fit for combat. He replied: "Yes, but we take very good care of them."

One day, a countess, accompanied by some French officers, came to the camp and asked me if I would let her nurses, from the hospital in Rennes, come to the camp and sell a few articles of lace to help the hospital. I could not refuse. The next day she returned and asked if I would also let the nurses sell some confections. I consented. Then she returned and said they had no flour or sugar and other things and asked if I would let her have them from the commissary to prepare the things to be sold. Of course I had to pay for what she needed. Her demands grew until I had to build a large house and arrange it for a bazaar. I sent the supplies to her, hauled the nurses and laces, clothing, food, and other things to the camp. It was a great affair and, besides making a large sum of money for the hospital, gave the men much pleasure. Many of the items were scraps basted to colored paper and trash, but the men were glad to buy them. The French-cooked pastries were especially welcome and sold for high prices. But seeing and talking to the attractive Red Cross girls was more valuable than money. It broke the strain and helped morale.

It was necessary to maintain discipline and soldierly deportment, but the intense occupation tended to make men lax about saluting. One day a soldier saluted me in an admirable manner. I asked his name and he said: "Private ——." I replied: "You are now Corporal —— because of your fine deportment and bearing." After that, there was little trouble about saluting.

My relations with the officers and men were very happy, and I looked forward to taking the brigade into action. I was much surprised, therefore, when an order came for me to proceed at once to take command of the First Artillery Brigade of the First Division.[16] I left late in the afternoon of December 20, 1917, and the regiments lined both sides of the road to bid me farewell as I passed between them. It was the hand of destiny, and a new phase of my career was begun.

This event was traced to my service with General Robert L. Bullard in the Philippines in 1899. I learned that General Bullard had been assigned to the command of the First Division and that the division would go into the line very soon. One of his first acts was to ask for my transfer. I proceed to Paris, where I spent one night, then went by rail to First Division headquarters at Gondrecourt. After reporting to General Bullard on December 22, I drove to the headquarters of the First Field Artillery Brigade at the Chateau de Beaupre. The historic, old stone mansion still bore the scars left by the Germans in 1871. It was not far from the lines, so we constantly heard the sound of guns in action. I joined just before Christmas and saw the American soldiers buy all they could from the commissaries and stores in order to give the French children the first Christmas that many of them had ever known.

I at once visited the regiments, consisting of the Fifth, Sixth, and Seventh Field Artillery, and inspected the batteries in their billets. The weather was very cold, and the men were suffering from exposure. The feet of officers and men would be so swollen that it was difficult to put shoes on in the morning. Gondrecourt was called the Valley Forge of the war. Although the troops had been trained by the French and had been in a quiet sector for a short time, they were rather inactive. To my surprise, I found one major and four captains absent without leave. It was necessary to tighten discipline and institute a vigorous program of training because we did not know what day we would move into the line. The troops responded wholeheartedly, and I soon felt confidence in our readiness.

A final maneuver of the whole division was held in a blizzard. We experienced great difficulty in moving the guns through deep snow and over the ice, but the guns were placed, and simulated fire was conducted in a gratifying manner. I found one officer using an old barrage table to support his infantry instead of computing a table for the problem. Thoroughness and accuracy were enforced in everything. The horses were on short rations, emaciated, and devoured by mange and vermin from the old French stables. Their care and conditioning became our first urgency.

Early in January 1918, the First Division received orders to relieve the First Moroccan Division of the French army north of Toul, which was directly in front of the Gondrecourt area. Reconnaissance and advance parties entered the sector to see the positions to be occupied and to become familiar with the missions to be assigned to our troops. The Germans had just delivered a heavy gas attack on the Moroccans, and the gun emplacements were infected with mustard gas.

On January 15, the first echelons of the division began the march. The weather was bitter cold, with deep snow and ice on the roads. Horses constantly fell, and progress was slow. A heavy rain turned into sleet. At night, the troops were billeted in villages. On January 18, the columns reached Menil-la-Tour, which became division headquarters and headquarters of the First Field Artillery Brigade. The First Infantry Brigade, consisting of the Sixteenth and Eighteenth Infantry Regiments, began to enter the line that night. The Second Brigade, composed of the Twenty-sixth and Twenty-eighth Infantry regiments, later extended the line to the right. The artillery began occupying battery positions by platoons on the night of January 22 so that some of the French batteries remained always ready to fire. The artillery relief was completed January 28. The Fifth Field Artillery with its 155-millimeter howitzers could cover the enemy's rear areas. The Sixth Field Artillery supported the First Brigade and the Seventh Field Artillery the Second Brigade, although all guns could cover any part of the division front. The First Trench Mortar Battery was posted at a sensitive point about six hundred yards from the German lines. The First Engineers constructed command posts, wire entanglements, and camouflage. The signal troops utilized existing lines but constructed a new system of communications. All other special troops were assigned to technical tasks. The two machine-gun battalions arrived in February and took turns occupying the lines. The field hospitals and ambulance companies took over French Adrian[17] barracks and soon had much to do. Artillery liaison officers were attached to infantry regiments. Sentinels were posted at the guns with a shell in readiness to be fired instantly at the point indicated by signal rockets.

The division was now in combat, for the Germans soon maintained considerable fire on the sector, both day and night, to which the artillery vigorously replied. Although the guns were camouflaged, all battery positions were well-known to the Germans. Because the position had been occupied since 1914, each side had at least a half-mile width of barbed-wire entanglements on their sides of the Rupt de Mad, a deep and narrow stream between the lines. The gun emplacements and the infantry trenches were deep in mud, while the dugouts where the men slept were cold and infested with vermin. Food and ammunition were taken to the troops at night. We soon had many sick and wounded, and the cemetery created for war dead grew rapidly. Enemy planes directed the fire of their batteries and also strafed our positions. Our air effort was feeble, and little help came from the French. I at once realized that all field artillery should be able to attack planes in order to carry out the mission of artillery to "destroy that element of the enemy which at the moment is most dangerous to the infantry."[18] I made that an objective the rest of my active service, but without success.

I visited all of the batteries daily and frequently went with one of the colonels to see the infantry front lines. Because my telephone and radio code name was Sitting Bull, the signalmen at each battery warned others when I left that "Sitting Bull is on the way."

A few days after occupying the sector, my eyes became so inflamed from contact with mustard gas in the preliminary reconnaissance that the doctor kept me in bed.[19] An orderly applied wet cloths to my eyes. At this time, another soldier reported to me as orderly. He told me he had sent my other orderly away. I asked him who detailed him. He said no one. He gave his name as Private William M. Steamer.[20] I kept him, and he proved an invaluable help and friend during the war and the rest of my life.

I soon found that the old arm prejudice of the infantry against the artillery was very strong. I endeavored in every way to remove it and to create a united spirit between the arms. Confidence was promoted by frequent exchanges of visits and by the instant response of the guns to a rocket signal or a telephone call from the trenches. The artillery had a number of casualties from enemy fire and from defective French ammunition, which exploded in the guns, killing the cannoneers. But morale was high, and pride grew with efficiency and confidence.

Both sides conducted raids to capture prisoners. While the artillery laid down a heavy box barrage covering the front on both sides of an area of the enemy's position, the raiding party rushed into the enclosed area and tried to find men caught in it who could not escape through the barrage. After our first raid, General Bullard wrote a letter commending the artillery. I told him that he and I knew of its efficiency but that I wanted the infantry to know it. This desire was soon realized, and the comradeship and devotion of the arms to each other became ideal.

On the night of March 3–4, my birthday, the division scheduled two raids. We rehearsed every feature of them, and our plans appeared to be perfect. The French gave me a large reinforcement of guns. At the time for the first raid, I fired an intense box barrage with 250 guns. After the scheduled time expired, the infantry reported that the raid had not taken place. We were puzzled but received no explanation. At the time for the second raid, I again fired a heavy box barrage, and the report came from the infantry that the raid was not made. General Bullard reported the failures to GHQ, which was waiting anxiously to hear the results. Soon, General Bullard received a call from GHQ asking who was responsible for the failure of the raids, no doubt for the purpose of making a victim of someone. The general replied: "I am in command of the First Division, and I am responsible for everything that takes place in it. If you want to fix responsibility for the failure of the raids, I am responsible." The effect was electric. Instantly, almost every officer and man in the division knew that he would never be a scapegoat if he did his duty. From that moment, there was complete mutual confidence in all grades and a determination that no one should ever fail. I believe that what became known as the "Spirit of the First Division" was born then and grew to perfection.

The next morning, we learned that while we had practiced every movement of the raids, we had not actually taken the Bangalore tubes[21] with which to blow the enemy's wire along our trenches to the front line. It was found that the tubes were too long to pass by the corners in the "zigzag" of the communication trench, and they could not be carried outside across our own wire. No one was to blame. Experience showed that only a small percentage of the raids were successful, and we need not have been surprised by our failure. No doubt but for General Bullard's courage and loyalty to his troops, GHQ would have crucified some innocent officer for not thinking about the length of the tubes. Then, as throughout the war, there was an impassable gulf between GHQ and the troops. GHQ knew little of the real conditions or difficulties and was always ready to sacrifice anyone who might be blamed. The incident illustrates the chances of success in anything when some slight feature of the undertaking was overlooked.

German planes bombed our lines and the division command posts at night. We had trenches near our billets and lost no time getting into them when we heard the planes coming. There were also many gas alarms sounded by bells in the villages. Everyone carried a gas mask and was required to put it on instantly. The saying was: "The quick or the dead." One day, Major George C. Marshall, operations officer (G-3) of the division, was passing through a village when the gas alarm sounded. He asked a sentinel if that was a real alarm or a practice alarm. The sentinel, busily putting on his mask and not noticing who was speaking, said: "Put on your gas mask, you damn fool, and don't ask questions." All enjoyed the incident but Marshall most of all.

Gas played a large part in the war and accounted for many of our casualties, but we fired more gas than the enemy. On one occasion, the Germans concentrated for a day and a night on the positions of a battalion of the Sixth Field Artillery. The batteries replied until all officers and men were overcome by the effects of the gas.

They were relieved at night by the rear echelon. But, because the emplacements could not be occupied and the guns could not be handled, we issued new equipment and occupied new positions. The disabled personnel were taken to the field hospital, where they lay in long rows on stretchers with their faces and burned bodies covered with bandages. None could speak, but all were proud and resolute. Under such intensity, the masks could not protect them. I was visiting the battalion when the gas concentration started but escaped injury. All of the battalion recovered. Our troops also repelled heavy raids by picked German troops who carried flamethrowers to burn our men in the dugouts, but a number of our infantry were killed in this way.

Our troops were commanded by the French generals.[22] We were visited by M. Clemenceau,[23] General Debeney,[24] the French corps commander, and other French generals.

During all this period, the question of discipline never arose. Both officers and men went beyond the call of duty from pride and devotion to the division and to one another. Many were cited for bravery. I tried to write to the families of our men who were killed, but soon the number became too great. Because all eagerly awaited mail from home, I often said that I would rather have a load of mail than a load of ammunition. People at home can never know the hunger of the soldier's heart for news and loving words when he is exposed to danger, hardship, and suffering. My dear wife's letters gave me strength and fortitude beyond estimate.

In order to identify officers and batteries, I ordered that when I approached a battery, the senior officer on duty should report to me by giving his name and identifying the battery. This they did in a most soldierly manner. At once, the men followed their example, and, as I reached a gun, the sentinel standing ready to fire would salute in a most soldierly way and say: "Private ——, Battery ——, —— Field Artillery reports, Sir." No matter how great the hardship and danger, the officers and men prided themselves on their military bearing and deportment. This was a part of the "Spirit of the First Division."

At one time, long sections of the wire of the artillery communication net would disappear during the night, and communications were cut off with the batteries. The most vigilant patrolling discovered no unauthorized personnel along the line. Then a signal corps officer attached to the brigade headquarters was sent elsewhere, and the sabotage ceased. He had a German name and occupied a room adjoining the artillery brigade staff mess. We recalled that he had always been in his room when we were eating. From the beginning, it was my rule during the war not to serve wine or talk about operations at meals. I never doubted that this man was a spy, but nothing could be proved.[25]

The artillery brigade staff was composed of superior officers and men. The intelligence reports, the tables of fire for each day, the liaison with the division and other headquarters, including the French, the munitions supply, the feeding and clothing of the troops, the care of materiel and horses, and the health of the troops were of the highest order within the limitations of procurement. My task was to coordinate, make decisions, and, through the regimental commanders, see that the batteries effectively executed the plans.

After the German surprise attack in Italy,[26] we adopted the elastic system of defense. Only a screen of men, mostly with machine guns, occupied the front trenches. A strong support line, well back, constituted the main line of resistance. A retired line of reserves was held for reinforcement or counterattack. All of the division artillery was withdrawn to sites where it could not be reached by enemy guns but could cover the support positions with all of its fire. Some 90-millimeter French guns were placed in advance positions to fire daily on the German trenches and batteries. We expected them and the forward screen of men to be lost in an attack, but they would delay and greatly punish the enemy.

When the great German offensive began on the British and French fronts on March 21, 1918, GHQ ordered the First Division to proceed to Picardy. The service in the Toul sector in Lorraine had cost the division in casualties 6 officers and 137 men killed, 9 officers and 384 men wounded, and 3 men missing, a total of 25 officers and 524 men. The small number of missing and prisoners was a tribute to the individual courage and fighting spirit of our men. In spite of the dangers and hardships, such a thing as neurosis never entered our minds.[27] Our soldiers were men. The army had not been forced to draft boys.

We were relieved by the Twenty-sixth Division, whose artillery I had known before the war at Tobyhanna and at Camp Coetquidan. Elements of this division began entering our lines on April 1. I remained to adjust the artillery brigade in its missions until April 5. The deep mud prevented our moving our 155-millimeter howitzers out of the sector or moving theirs into position. We exchanged guns and materiel so that the incoming division manned our howitzers and we took over theirs where the regiment had halted in the road. Thus was realized my prediction as to using such heavy materiel for divisional artillery. General Lassiter, commanding the artillery brigade of the Twenty-sixth Division, and I were good friends, and the relief was executed entirely to his satisfaction. He had relieved me in command of the Sixty-seventh Field Artillery Brigade at Camp Coetquidan.

Later, when we were intensely engaged in Picardy, the inspector general of the First Corps investigated a malicious report against General Bullard and me falsely charging negligence in the relief and found it to be entirely false. The investigation was inspired by the chief of staff of the First Corps,[28] who had a sadistic mind for injuring others. I had been a close friend of his parents and had known him since he was a boy. Afterward, he had General Edwards[29] relieved from command of the Twenty-sixth Division on false charges and blighted the career of an able officer. General Bullard remarked that while we were fighting desperately with our faces to the enemy, our own command was stabbing us in the back. We never felt safe from spying and false accusations, inspired by jealousy on the part of the men we were serving. Later, the First Division and I were to drink the dregs of this bitter cup when the division obeyed the order to capture Sedan.

The division was moved by rail from Toul to the Gisors area, where it was prepared to enter the Montdidier sector, seventy-five miles away. While in the Toul sector, little forage had been provided for the horses; thus they were thin and emaciated. The artillery brigade was inspected by General Micheler[30] of the French army. I had it formed on a slight slope so that in moving away after the inspection the horses could pull the carriages in the presence of the inspector. They were a pitiful sight. General Micheler said to General Bullard: "Those horses cannot maneuver on the field of battle." General Bullard replied: "They are Texas horses, and Texas horses are thin." When the command was given to march, a platoon commander tried to turn his horse around to give his command when the horse fell and could not get up. General Micheler looked at him sadly and rode away.

General Pershing visited the division and talked to the officers. All felt the responsibility of taking over an active sector still reacting to the German attack. On April 17, the division began the march to the front, and the sound of the guns in the lines warned us of what awaited. The horses could scarcely pull the guns. All officers and men were required to walk. After four days, the organizations took up billets and sent advance parties to begin the relief of a French corps in the Le Mesnil–St. Firmin sector near Montdidier.

The batteries soon relieved the French guns and took over their missions of fire. The German guns were very active, and our daily tables of fire included all enemy battery positions, infantry trenches, and communicating roads. The horses of one of our batteries were placed in an old French stable well back of the line. Almost immediately, the German heavy guns concentrated on it one night and killed all of the animals. Even when we took the horses to water in the rear area, the Germans would strafe them. No movement to the front could take place during daylight, so food and ammunition were carried to the troops at night and the killed and wounded evacuated. The position we occupied roughly outlined where the French had made a stand when driven back by the Germans, which required us to dig trenches and camouflage gun positions. At one time, all of the guns in one of our batteries were disabled in a few minutes by concentrated artillery fire. Some of our advanced billets were wrecked and many men killed by artillery fire. After a destructive fire at Villers Tournelle, I visited the infantry regimental headquarters in a deep dugout in the town. Shells were bursting frequently. Mangled bodies were in the streets. When I left, the colonel asked me to take his French liaison officer with me. I did not want to run, but the bursts were all about us. The French officer exclaimed: "Let's run, General. This is no place for us." I agreed with him. At another time, I was to take dinner with an infantry brigade commander when a heavy concentration began on his command post. I telephoned that I did not like the way he planned to receive me and would visit some other time.

The division and artillery brigade headquarters were in a badly battered house in Le Mesnil–St. Firmin. One night, a shell struck the corner of the house under my room, completely wrecking it. Other parts suffered similarly. After that, we slept in the wine cellar. Here, we were visited by M. Clemenceau and General Mangin[31] and by a number of officers from Washington. One day, Mr. Bernard M. Baruch[32] and Mr. Dwight W. Morrow[33] drove up in the afternoon, and I gave them lunch in our mess.

Our casualties in killed and wounded mounted rapidly. All billets in advanced villages were abandoned, and the men sheltered
themselves in trenches. It became known that the Germans were
about to renew the attack in this sector.

At once, a counteroffensive was organized for us to capture
Cantigny, a key position.[34] The Twenty-eighth Infantry trained and
rehearsed to make the assault, assisted on its flanks by the other
regiments. The task of the artillery was to neutralize the enemy
by a heavy preliminary preparation and to fire heavy barrages in
front of the advancing infantry. The French reinforced our artillery
by 132 75-millimeter guns, 36 155-millimeter guns, 178 heavy guns
and howitzers, and 40 trench mortars. These, added to our brigade,
made 564 guns. Positions were occupied. Tables of fire for the operation were calculated for each position. All batteries adjusted by
firing a few shots. The attack was to begin at 5:45 A.M., May 28. At
4:30 A.M., May 27, the enemy delivered a terrific preparation fire and
at 6:00 A.M. sent over a strong raiding party. We lost no prisoners and
captured three Germans, but we had a number of casualties, especially in the engineers. At 5:45 A.M. on the twenty-eighth, every gun
opened simultaneously on each assigned target. This silenced the
enemy guns and pinned his infantry in their trenches. French observation planes reported German batteries not previously located
and directed the fire to silence them. At 6:45 A.M., the 75-millimeter
guns began the rolling barrage, and the infantry followed close to
the bursting shells. During the preliminary bombardment, French
tanks passed to the front lines and accompanied the infantry. Flamethrowers were used to drive the Germans out of their dugouts. The
entire objective was taken by 7:30 A.M., on schedule. We pushed patrols forward, and the infantry consolidated their gains. The artillery enclosed the captured area in a great box barrage. We had killed
many Germans and captured 5 officers and 225 men.

The French withdrew all of their reinforcing guns, leaving the
seventy-two guns of the artillery brigade to protect the position and
help repel counterattacks. The enemy soon reacted by shelling the
position heavily and making repeated counterattacks, which we re-
pulsed. We were aided by French planes. On the night of May 30–31,
the Twenty-eighth Infantry was relieved by the Sixteenth Infantry.
At that time, the total American losses were 45 officers and 1,022
men. We had only 1 officer and 15 men missing, who might have
been prisoners. The success of the battle raised Allied morale and

gave the American army a place of importance. It was respected alike by the Allies and the Germans. The various high commands praised and cited the Twenty-eighth Infantry.

After the battle, General Pershing came to the division and expressed great satisfaction. He asked me to walk with him away from the others and said: "Summerall, I am going to make you a major general, and I want you to learn to handle infantry as well as you handle artillery." I replied: "General, all I know about artillery I learned from the study of infantry and how artillery could serve it." We returned to the house, and he said little more. I felt vindicated in my stand at his headquarters in Paris when he told be that he wanted me to get together with his staff in my controversy about artillery. Here, as always, we did not have enough guns, and the infantry paid the price in casualties. I recalled also my paper at West Point for the Thayer Club on artillery, which the *Infantry Journal*[35] wanted to publish because the infantry said it was all about infantry.

After Cantigny, the Picardy sector became intense. Our front lines were thinned and echeloned. Batteries were withdrawn to defend in depth. Command posts were retired. The division and artillery brigade occupied Tartigny. Enemy aviation became aggressive, and bombing killed the division quartermaster[36] and others in their billets far to the rear.

We received information that the German general offensive would soon be resumed. All troops were ordered to fight in place. Early on June 9, an intense bombardment with shell and gas began on our entire positions and rear areas, which continued through June 13. Enemy infantry attacked the French on our right, who fell back, leaving our flank exposed. We never knew why an assault was not made on us, but it may well have been considered too costly after Cantigny. When we learned that the French were withdrawing, General Bullard and I went to General Debeney, the army commander, and asked him if he did not intend for the troops to hold their positions. He said yes but looked hopeless as to what the French were doing. General Bullard told him that we intended to hold our lines, which pleased him greatly.

The greater the hardship and danger, the higher the morale in the division. Captain A. B. Butler, my aide, prepared a cartoon daily that was mimeographed with news of the war and sent to the trenches and battery positions every night. The men looked forward eagerly to receiving it.

The spring made Picardy a field of beautiful poppies. Men filled empty shell cans with them and decorated their camouflaged gun emplacements. They found scraps of furniture and rugs in the destroyed and abandoned villages and dressed the space around the guns. Deep dugouts were prepared for cover during the heavy bombardments. The fields were covered with grass, and our half-fed horses were grazed during the night. Creolin[37] was obtained, and the lice and mange were destroyed. The luscious grass soon made flesh to cover their bones. The men jestingly said that the horses were groomed all day, grazed all night, and had no rest. The space around the kitchens was covered with white sand and kept immaculate. The small tree limbs were laced over the paths. The latrines and stables were kept in the most sanitary condition. Carriages were painted, and the men were clean and well dressed. It became difficult for me to make my inspection rounds as the men insisted on showing me all they had done. I praised them personally and published orders of commendation when mange and lice disappeared from a battery. They wanted no greater reward.

On the Fourth of July, we had a horse and transportation show in the woods adjoining the chateau at Grivesnes. It was unbelievable to see the fat, well-groomed horses and mules with glossy coats and the beautifully painted caissons and wagons. On approaching one wagon, I saw the driver standing by the wheel with a puzzled expression on his face. He did not want to step on the double tree or wheel to mount to his seat for fear of scarring the paint, and he said: "How in hell am I going to get up there?" The gun emplacements also competed on the condition of the materiel and the decoration of the emplacements. The judges made many awards, and the winners were given money and a three-day leave in Paris. Prizes were awarded and leaves and money given to the winners. In spite of danger and hardship, there was great cheerfulness.

At the same time, a battalion of infantry was sent to Paris to take part in the Fourth of July parade. At noon on July 4, a salute to the Union was fired by the 155-millimeter guns of the Fifth Artillery. Forty-eight shells were fired into the most sensitive parts of the enemy's lines. That day, orders were received for the First Division to be relieved in the sector by two French divisions. The troops moved out on the nights of July 5–6 and 7–8 and went into billets in the Beauvais area. During the occupancy of the sector, our losses amounted to 64 officers and 958 men killed, 88 officers and 3,809 men wounded, and 9 men missing, or a total of 152 officers and 4,776 men, showing that the severe fighting was constant. Replacements arrived, and the division began training for open warfare, to which we had long looked forward. It was not long, however, before the great German attack on the Marne Salient, which began May 27, forced alteration of our plans. On the night of July 14, the Germans crossed the Marne River, driving the French before them, and were advancing rapidly, with little opposition, on Paris. In the meantime, the First Division, under the Tenth French Army, had begun moving to the threatened area. On the night of July 11, horses, guns, and men were hauled by trucks, except for the heavy guns of the Fifth Field Artillery, which were towed by trucks. The movement was continuous day and night in anticipation of the German advance.

Shortly after the relief in the sector, General Bullard was assigned to command the newly organized Third Corps. He at once sent me to GHQ to arrange for assignment to the artillery of his corps. En route, I stopped at the headquarters of the First American Corps and found an order promoting me to major general. I at once took the oath of office. At GHQ, I was offered the assignment as chief of artillery of the First Army. I stated that I wanted to be assigned to the command of the First Division, and this was approved. I hastened back to join the division and assumed command July 13.[38] Orders arrived for the division to enter the battle line and attack in the direction of Soissons on the morning of July 18. The positions were reconnoitered, a conference was held, and orders were issued on July 17.

Chapter 17

Soissons—the Decisive Battle of the War

There can be no doubt that during the Battle of Soissons the heroic advance of the First Division, in spite of unprecedented losses and the most determined resistance of elements of eight German divisions, turned the fate of the war.[1] The First Moroccan Division on its right had to be relieved after twenty-four hours, while the Second American Division on the Moroccans' right lasted less than two days. The First fought desperately from July 18 until July 22. It cut the enemy's road and rail communications, besides killing and capturing large numbers. On the night of July 20, the Germans recrossed the Marne and began the retreat that never ended until the Armistice.[2]

I visited the lines each evening when the advance stabilized. On the second day, I was informed that General Pershing and the French army commander[3] would visit the division headquarters during the afternoon.[4] I left early for the lines and told the chief of staff[5] to tell them I was compelled to see the troops. One brigade had not reached its objective. On leaving the division command post, I met the lieutenant colonel of a regiment in this brigade.[6] He told me that he was taking supplies to the regiment. I told him to come with me. His brigade had not reported taking the objectives. On reaching the brigade command post, I found the brigade commander[7] much confused and worn. I told him to get some rest, that the attack would be resumed the next morning, and that he would lead the attack. On reaching a regimental command post, I found the colonel[8] exhausted. He was sullen and defiant. I asked him why his regiment had not attacked. He replied: "The order was impossible, and I did not try to obey it." I could have relieved him, but it was evident that he was overwrought and scarcely responsible. The strain had been too great for him. I told him that I had brought the lieutenant colonel, who would be on duty while he rested. The colonel was killed the second day after this.

I took a young captain to guide me to the lines. On the way, a heavy German artillery fire began, and we took cover in a shell hole. When I asked his name and where he came from, I found that he was the son of one of my teachers at Dr. Porter's. Despite the shelling, I found that the men were in good spirits. I explained the situation and the necessity of taking the objective of Berzy-le-Sec in the attack next morning. They assured me that they would take it if I would give them our own artillery support because they did not like the French artillery. I told them I would do so. It was here that I first adopted the method of advancing in two or three successive echelons, leading each echelon with a barrage from all of our guns.[9] I had reported at the Paris conference[10] the number of guns the French and British believed were necessary for attack, to which the GHQ staff took violent exception. Here, I found the truth of my contention and multiplied my guns by diminishing the front covered by their fire. This method was used in all my subsequent attacks by the First Division.

The next morning, the advance was successful, but the lieutenant colonel whom I took to the lines was one of the first men killed.[11] Years later, I was told that my enemies at GHQ criticized our losses and that General Pershing came to investigate. When he found that I had gone to the lines, nothing further was said. None of them (the GHQ crowd) were ever seen in the lines.

On the third afternoon, when I reached a regimental command post, the colonel[12] at once said in a most resentful manner: "General, my regiment has lost 60 percent of its officers, nearly all of its old noncommissioned officers, and most of its men, and I don't think that is the way to treat a regiment." I could have relieved him for insubordination, but I replied calmly, telling him that the tide of battle had been changed by his troops and those of the division of which his were only a part and that the attack would be resumed with what he had. In repeating the incident later, he stated that I replied: "Colonel, I did not come here to have you criticize my orders or to tell me of your losses. I know them as well as you do. I came to tell you that the Germans recrossed the Marne last night and are in full retreat and you will attack tomorrow morning at 4:30." He added: "From that day, I have never questioned your orders, and I never will." He was about the best combat officer I ever knew.

At another time, the division was to attack by orders of the French corps at 4:30 P.M. At 4:25, the colonel of a regiment[13] called me and said that he could not obey the order and would not attack. I told him as calmly as possible that he must not say anything to his battalion commanders, whom I knew would attack. He showed himself unsuited to be a combat regimental commander. The battalion commanders led their troops to take their objectives. I found that the colonel's connections were such that it would be best not to relieve him, but after the battle he was transferred out of the division. Thus, the two colonels in a brigade stated that they could not obey an attack order, and the brigade commander was too worn and mentally confused to force the attack. Such was the terrible ordeal of battle on officers.

While I had not anticipated such conditions, my going to the lines discovering and correcting them and heartening the men in contact with the enemy by talking to them, praising them, and explaining the results of our success enabled me to influence the battle successfully. I tried to see the men in combat and their officers as often as possible. The confidence and devotion that I received added to the reward of victory.

On the first day of the battle, the commanding general of the First Moroccan Division[14] on our right called the chief of staff and said that they had taken three hundred prisoners. As we knew that the Moroccans killed their prisoners, the chief of staff[15] replied: "Beaucoup pour vous, mon general." A little later, the general called again and asked us to give him three hundred prisoners, explaining that his had disappeared and that, as he had reported them, he had to turn in three hundred. After having his promise to keep them, we gave him three hundred prisoners.

The division attacked in line of regiments, each regiment in column of battalions, designated from front to rear as assault, support, and reserve battalions. The interior reserve battalions were at the disposal of the brigade commanders, and the flank reserve battalions were kept for employment by the division commander. The French division on our left never tried to advance and left our flank entirely exposed. I had used the division reserve battalion on the left and found that, on account of losses, I must take the reserve battalion from the right regiment. To my amazement, I found that the regimental commander[16] had used it without authority. It was necessary for me to take a weakened battalion. I did nothing to the regimental commander, though his action was inexcusable. He had acted in good faith to meet what he considered an emergency.

The chief of staff of the division, Colonel Campbell King, had been a boy with me at Dr. Porter's. I frequently marveled at the fate that had charged two Porter boys with the responsibility for this great battle. He was a superb officer and the greatest help and comfort to me.

On one of my visits to the lines, I found a captain far to the rear away from his company. He was, of course, overcome by the danger and was avoiding it. I asked his name and where he came from and then told him to think how his family would feel if they knew what he was doing. As far as I knew, he behaved well after that.

At one point, I found a company well deployed in front and depth, with the men sheltered in foxholes. On approaching, I asked what officer was in command. A voice replied that they had lost all of their officers. I asked what noncommissioned officer was in command, and the same man replied that they had lost all of their noncommissioned officers. I then asked who was in command, and the man rose from his foxhole and said: "I guess I am, Sir." I asked his name, and he said "Private ——." I said: "You are now Sergeant ——." And I praised him for his action in taking command and organizing the position. It could not have been better.

I found a rolling kitchen one night serving soup to the men in foxholes close to the enemy. As I approached, the cook called out: "General, I am Hatfield. I was your cook in Battery B at Fort Myer." I shook his hand and told him I was glad to see him. It was by seeing the men in their danger and talking to them that I gained their confidence and made them feel that they were fighting for me and that I shared their danger and understood them. They were always buoyant and confident and inspired me to be certain of success.

But one day, on my way to the lines, I picked up two men who were fresh with their guns and packs after two days of battle. I saw at once that they had been skulking and were coming in for food. Because they claimed that they had been lost, I took them to their regimental commander and told him to send them at once back to the front lines. The men there would see that they did not skulk.

These cases of neurosis of a few officers and men show what was expected and how I dealt with them. Later, I had similar trouble in the corps that I commanded. The great mass of officers and men will overcome their fears, and the further they are to the front, the less they fear. At one time, the division judge advocate[17] brought to me five sets of charges against men for cowardice. I told him to drop the charges and send the men to their colonels, who would at once place them in the front lines, where their fears would disappear or the other men would take care of them. The difficulty disappeared as the men had more experience in battle. I have said that failure in battle results not so much from the resistance of the enemy as from a state of mind of officers and men that deprives them of a desire to fight. By personal contact and assurance, a leader must inculcate and maintain a will to win in every member of his command. I never tried a man for cowardice.

On all of my trips to the front lines, I passed hundreds of bodies of our men and greater numbers of bodies of Germans. The sight was heartbreaking, but I knew that we were winning a great victory and deciding the fate of the war. It was necessary to pay the price, but the sacrifice was compensated for by the results. I felt that my life and that of the troops must be given if necessary to gain the final victory. We could not bury the dead because the losses among the burial parties by the enemy fire were too great. I resolved to do something to immortalize their names and their triumphs. The monuments on the battlefields of France and in Washington were the fulfillment of this vow to them. When the enemy had withdrawn, their casualties were buried by the Fifteenth Scottish Division, who sent the personal articles that were recovered from the bodies of the dead to us for later shipment to the families of those who were killed.

The army had notified us that we would be relieved by the Fifteenth Scottish Division on the night of July 21, which I told the men on my visit to them that day. When we learned later that the Fifteenth Scottish Division could not arrive until the next day, I sent my French aide to the French corps commander[18] to find out why we were not relieved as promised. When he explained, he asked the aide what the First Division intended to do, as though we would march out. The aide saluted and replied: "General, this is the First American Division." I never again made any promises.

The Fifteenth Scottish Division, under the command of Major General H. L. Reed,[19] began to arrive during the morning of July 22. General Reed and I, with a liaison officer from one of the brigades, Captain J. A. Edgar, went along the lines during the afternoon. At one point, we came under heavy strafing by German artillery. I did not want to show any fear of it, but we were in great danger. Soon, General Reed said: "General, let's double time." I agreed, and we ran for cover. We were surprised by the field equipment of the Scottish division. In battle, we carried nothing but our mess kits and blanket rolls. We did not even think of digging a latrine. But the British had complete mess equipment and bedding and set up a most complete latrine with tent. I had occasion to use it when the orderly said: "This for the general officer commanding." As I was still in command, I replied: "Well, I am the general officer commanding."

Our orders required the infantry to withdraw during the night as soon as they were relieved by the British infantry. As the British were to attack early in the morning of July 23, our artillery brigade remained to cover the assault. The First Sanitary Train[20] also remained to care for the British wounded. The troops assembled in a woods where a kitchen marked the place for each company. Many men had been separated from their units, and here all were again united. A number of men from other divisions had become separated and joined the First Division. We served them a hot breakfast. I at once visited every company, praising their valor, and telling them of their success and what it meant. All were pitifully worn, hollow cheeked, dazed, and exhausted. Practically every man fell sound asleep, for they had had no rest and little food for over four days. The companies were pitifully reduced in numbers. All companies were mere skeletons of the well-filled ranks that entered the battle. Some companies were commanded by privates. One regiment was commanded by a very young captain.[21] Every battalion commander was a casualty. After the men had eaten, the bands played to cheer them.

In this battle, the division fought continuously for five days, the artillery for six and seven days. The division gained eleven kilometers against some of the best German divisions. It penetrated the angle of the salient made by the German army in the attacks beginning in May and culminating in the second battle of the Marne and the advance on Paris, July 14. We cut his rail and road communications and were chiefly responsible for his recrossing the Marne July 20 and beginning the retreat that never ended until the Armistice. So powerfully did the division fight that the Germans never captured a prisoner or crossed our lines to identify our dead. Six weeks after the battle, the German order of battle showed that the First Division was still in Picardy. The division captured 125 officers, 3,375 men, 75 field guns, 50 trench mortars, 600 machine guns, and quantities of rifles, ammunition, and stores. Enemy dead covered the field.

The division lost 77 officers and 1,637 men killed, 157 officers and 5,335 men wounded, and 111 missing. This was a terrible price to pay, but we knew to the last man that the war would be won as a result of this decisive battle. The division had only its seventy-two guns for nearly three miles of front. A regiment of French portee 75-millimeter guns[22] was useless as it remained too far to the rear and soon left. Our heavy losses were directly due to our lack of sufficient artillery to neutralize the enemy, as I had warned the staff in Paris. Those men really knew nothing about firepower in battle or the meaning of artillery. Corresponding results from insufficient artillery occurred in every battle. The artillery of the division at Soissons and elsewhere was superb in its quick concentrations of fire where needed, in its accurate and dense barrages, and in its displacement forward as the lines advanced to support the infantry closely. General Mangin, the commander of the Tenth French Army under whom we fought, published an order of commendation. Here was developed what in World War II was called a *combat team*, by uniting infantry and artillery as a unit.

As soon as possible, I held a critique of the battle with the officers. While no serious mistakes were made, a number of details required improving. When it was over, a brigade commander asked me where I learned so much about infantry. I replied to him as I did to General Pershing that knowledge of infantry is necessary to learn about any other arm. I learned artillery by studying infantry.

Chapter 18

Recovery

As much as it had suffered, the First Division had little rest. Beginning July 23, French trucks conveyed the foot troops as rapidly as possible to the Saizerais sector in Lorraine on the left of the Moselle River, where they relieved the Second Moroccan Division during the last days of July and the early part of August. The artillery followed. The sector covered a front of eight kilometers and had successive rows of wire and trenches. We took up the elastic system of defense. Battalions were rotated in the three lines, and training was conducted for those in the support and reserve positions. Some French ninety-millimeter guns were used forward for harassing fire. I had told the troops that men were never held up by machine-gun fire unless they wanted to be held up. During a division attack training problem, I noted a squad halted in front of a simulated enemy machine gun. I asked the corporal what he was doing. He replied: "We are overcoming machine-gun resistance."

On August 9, some Japanese staff officers visited the division, and I took them through the successive lines of trenches. As we passed along the most forward trench with only a machine-gun squad at intervals, one of them remarked with some superiority that we were not holding the line. I told him no but did not enlighten him on the elastic system. They were very patronizing. While it was supposed to be a quiet sector, we were subject to frequent bombardments and several raids in which a few prisoners were taken and men killed and wounded. Here were received seven thousand replacements of officers and men, who underwent training. The new men soon imbibed the spirit of the division. For relaxation, we granted short leaves to officers and men, usually for visiting in the sector. I generally had officers to lunch and constantly visited commanders at their headquarters and the men in the trenches.[1] We deloused and issued fresh uniforms. Morale was very high.

The high command was preparing the next battle, which was

to be the reduction of the St. Mihiel Salient, which the Germans had held for four years. During the night of August 23–24, the First Division was relieved by the Ninetieth Division and proceeded to its training area at Vaucouleurs for the battle. The losses in the sector were one officer and fourteen men killed, one officer and thirty-one men wounded, and four men missing.

At Vaucouleurs, the attack on the St. Mihiel Salient was rehearsed on similar terrain during the week of training and equipping. I gave a lunch for the Red Cross workers and met there for the first time Colonel Clark Williams,[2] who became my closest friend and did much for the division. I expected many losses, and on the days preceding the assault I acquainted him with the plan of attack and asked him to ensure Red Cross help to care for the wounded. His accomplishments were all that could have been desired.

Chapter 19

The St. Mihiel Salient

The front assigned to the First Division for the attack was the old Toul sector that it had first held. The terrain was familiar. The division assembled in the Forêt de la Reine, and units were rehearsed in their parts. Our objective was the angle of the salient, just as it had been at Soissons. We were to penetrate the enemy's position and meet troops attacking from the opposite angle of the salient, thus cutting it off. It was necessary for us to pass through at least half a mile of our wire, cross the unfordable Rupt de Mad, and pass another half mile of enemy wire, bey ond which was the Madine Creek, protected by wire. The task was very formidable. The high hill on our left, called Montsec, was heavily fortified and commanded a view of our area.[1]

The plan of attack was for the engineers to proceed with Bangalore tubes to blow lanes in our wire. They were followed by engineers carrying floating bridges to be shoved over the Rupt de Mad for the infantry to cross and capture the German advanced line. Then the engineers were to blow the German wire, and the infantry were to cross it and occupy the position. The advance was to continue until it met the troops from the north. The French on our left were not to attack. We were to have a regiment of infantry face to our left in front of Montsec to protect our left flank. The attack was to begin at 5:00 A.M., September 12. An artillery neutralization fire from 168 guns was to begin at 1:00 A.M. and change to a rolling barrage in front of the infantry at 5:00 A.M.

At a conference before the attack, I pointed out the difficulties and ordered that the infantry should swim the stream if the bridges failed to arrive. Lieutenant Colonel George S. Patton, who commanded the tank battalion, said that he would drive the tanks into the stream and the infantry could cross on them. I replied that the men in the tanks would be drowned. He answered: "Yes, and I will be one of them." I told him that I could not resort to such sacrifice. Fortunately, we found that the stream was fed by a lake some miles to our left. After we cut the dam the night preceding the assault, the water became low enough in the stream for our men to ford it.

We were operating in the First Corps, commanded by General Joseph T. Dickman.[2] He was ignorant, foulmouthed, and brutal at the conference held with division commanders. On the night of the attack, a vicious member of the GHQ staff[3] came to my headquarters with every evidence of watching and spying on me. I was sure that he was looking for some act or word of mine to report and have me relieved.

The attack went off as planned in every detail. The engineers with the bridges waded the stream and continued forward looking for a stream to bridge. Finally, one man exclaimed: "Where in hell is that river that we were to bridge?" The wire cutters and the bridge details took arms from the killed and wounded and advanced with the infantry until the objective was taken. The division had a squadron of cavalry attached, with the idea of charging after the wire was crossed, but the terrain did not permit it, and the machine guns of the enemy would have made it too costly. The cavalrymen did some courier service, but the day of cavalry on the field of battle had ended. The fellow Dickman tried to discredit me for not having it charge. He knew nothing of conditions, and he was totally ignorant of battle. All objectives were taken on schedule, and our losses were unbelievably small.

The division on our right (the Forty-second) bore away and created a gap of several miles. Portions of the Third Division were attached to the First to cover it. A division sent its trains over our communicating road, thus depriving us of its use and causing great inconvenience. We adopted other means, and I never complained. It was a rule of the First Division never to criticize other troops. I always felt that I was fighting Germans, not Americans, and I never criticized other troops as other commanders did.

There was no doubt that this battle was made less difficult by the victory at Soissons. The enemy's will was broken. The enemy set fire to all of the villages in his retreat. He saw that the war must be lost. Our losses were 3 officers and 90 men killed, 10 officers and 430 men wounded, and 10 men missing. The division headquarters moved forward the next day, and a new line was formed close to the retreating enemy, where we halted. A conference of officers was held to discuss the battle and profit by our experience. General Pershing visited the division and congratulated us. On Sunday, the chaplains conducted services and gave thanks for our victory and small suffering. All felt that the Lord was with us.

Chapter 19

Chapter 20

The Second Phase of the Meuse-Argonne

On the night of September 19, the foot troops moved by trucks and the mounted troops by marching to the area near Verdun and were placed in reserve in the Third Corps. The First Division received orders to attack a very strong Austrian division in front of Verdun, and we prepared accordingly. The great Meuse-Argonne offensive of the First American Army began early on September 26. On September 27, just as the First Division on the right of the line was about to execute the assault in front of Verdun, it was transferred to the First Corps on the left of the line. The division proceeded across the rear of the army in two night marches. It then transferred to the Fifth Corps and relieved the Thirty-fifth Division on the night of September 30. This division was badly disorganized, with no defined line.[1]

The First Division advanced with regiments in line and in column of battalions until the leading troops came under enemy machine-gun fire. The different echelons then occupied positions for attack. The artillery arrived the following night. During the night advance of the infantry, a staff officer asked to which division and corps the troops belonged. A voice from the ranks replied: "We don't belong to any corps. We go where we are needed most." Although the First Division had received orders to attack at once, the orders were countermanded, and the troops remained for three days under heavy enemy fire, during which time we lost five hundred men and officers. The terrain, broken by deep ravines and woods with high hills and knolls, gave the enemy commanding positions. Patrols sent out by orders from the corps sustained heavy loss and exposed us to the possibility of having the enemy learn our intentions. Because we knew the enemy's dispositions from the contact of our lines, patrols were needless.

At 5:25 A.M., October 4, without any preliminary fire, the barrage dropped two hundred yards in front of our jumping-off line and stood until the infantry advanced to the line of the bursts. It then moved forward, closely followed by the infantry. When resistance was intense, our men passed over the ground from which the Thirty-fifth Division had been driven and where many dead still lay on the field. The left brigade gained the Exermont Ravine and Fléville, which the troops were ordered not to occupy. Enemy artillery intensely bombarded the area, but this was the only spot on the Western Front where we gained the objective that day. The troops dug in at once. The right brigade met very heavy resistance, advanced two kilometers, but did not gain its objective. The left brigade advanced four kilometers. Because the troops on our flanks did not advance at all, we were exposed to flank fire. Forty-seven French tanks, distributed along the line, gave great assistance until most were disabled.

On October 5, the first objective was to bring the right brigade in line with the left. All remaining tanks accompanied the right brigade. At 6:15 A.M., divisional and corps artillery fired in front of the right brigade, using many smoke shells to screen the troops. The barrage advanced at 6:30 A.M., closely followed by the infantry. The exposed right flank suffered from enemy fire. Although the fighting was desperate on both sides and the losses heavy, the line was gained on schedule. The barrage then fell along the entire line to the second objective. This was gained and the line extended through the Bois de Moncy. A patrol discovered that Hill 269, several hundred yards to our right and in the rear of the enemy's line facing the division on our right, was unoccupied. A detachment was sent to occupy it, but this was driven away after several hours. Then a battalion of engineers was sent to recapture it and was, in turn, driven off. But it counterattacked and held the hill. On October 6, in conformity with corps orders, the division organized its position and sent out patrols. During the night of October 6, the Eighty-second Division moved into the ground occupied by the First Division and at daylight of October 7, faced to the left flank and attacked the north of the Argonne Forest, crossing the Aire River Valley. The guns of the First Division turned to the left and supported the attack.[2]

The First Division was now transferred from the First Corps to the Fifth Corps. The occupation of Hill 269 had forced the Germans to withdraw beyond that hill. Our zone of action was extended to the hill, amounting to at least a brigade front. The left flank was also extended to the Fléville-Sommerance Road. On October 8, two German counterattacks were driven back. Orders were received from

(Above) Officers of the Fourth and Fifth Artillery in the Philippines. Lieutenant Albert J. Bowley, Lieutenant John W. Kilbreth Jr., Captain Sydney W. Taylor, Captain Henry J. Reilly, Lieutenant Miller, Charles P. Summerall, Lieutenant Marcus McCloskey. Courtesy of the First Division Museum at Cantigny. *(Below)* Field gun being transported by river in the Philippines, 1900. Courtesy of the National Archives.

Captain Reilly's battery at drill, Peking, 1900. Courtesy of the National Archives.

The Forbidden City, Peking, 1900. Courtesy of the National Archives.

Skagway, Alaska, and Lynn Canal. Courtesy of the National Archives.

Reception at the West Point superintendent's house, 1905. Courtesy of the National Archives.

U.S. Military Academy cadets conducting a coast artillery drill. Courtesy of Charles P. Summerall Papers, box 34, Library of Congress.

Troops on maneuver at Tobyhanna, PA, field artillery range. Courtesy of Charles P. Summerall Papers, box 34, Library of Congress.

Troops standing inspection at Tobyhanna, PA, field artillery range. Courtesy of Charles P. Summerall Papers, box 34, Library of Congress.

Members of the Chauncey Baker Mission and American Expeditionary Forces General Headquarters in Paris, July 1917. *Right to left, front row:* Colonel Frank Parker, Colonel Robert H. Dunlap, Captain James L. Collins, Major Hugh A. Drum, Captain William O. Reed, Captain Martin Shallenberger, Colonel Dennis E. Nolan, Colonel John McPalmer, Colonel Kirby Walker, Captain Nelson E. Margetts. *Right to left, back row:* Major Marlborough Churchill, Colonel Edward Anderson, Colonel Sherwood A. Cheney, Colonel Mark L. Hersey, Colonel James G. Harbord, Colonel Hanson E. Ely, Colonel Chauncey Baker, Colonel A. L. Conger, General John J. Pershing, Colonel William S. Graves, Colonel Fox Connor, Colonel Dwight E. Aultman, Major Alvin B. Barber, Lieutenant Colonel John W. Barker, Colonel Charles P. Summerall, Colonel George S. Simonds, Colonel Robert Bacon, Colonel Benjamin Alvord, Major Morris W. Locke. Courtesy of the National Archives.

(Above) Summerall at Fifth Field Artillery mess, July 4, 1918, Tartigny, France. Courtesy of the National Archives. *(Below)* Summerall and Pershing at an awards presentation, September 7, 1918, Vertuzey, France. The officers in the second row are John Quekemeyer (Pershing's aide), Alban Butler (Summerall's aide), and Frank Parker (then commanding the First Brigade, First Division). Courtesy of the National Archives.

(*Above*) Summerall visiting the troops, 314th Engineer Regiment, Eighty-ninth Division, November 13, 1918, at Pouilly, France. Courtesy of the National Archives. (*Below*) Major General Charles P. Summerall, commanding general of the Fifth Army Corps, with his staff and commanders. *Seated, left to right:* Major General George B. Duncan, commanding general of the Eighty-second Division: Major General Charles G. Mortong, commanding general of the Twenty-ninth Division; Major General Charles P. Summerall, commanding general of the Fifth Army Corps; Major General Harry Hale, commanding general of the Twenty-sixth Division; Brigader General Wilson B. Burtt, chief of staff, Fifth Army Corps. *Standing, left to right:* A. W. Foreman, assistant chief of staff, G-1, Fifth Army Corps; Colonel G. M. Russell, G-2; Colonel Gordon Johnston, Eighty-second Division; Colonel T. H. Emerson, G-3, Fifth Army Corps; Colonel Sidney A. Cloman, Twenty-ninth Division; Colonel Duncan K. Major Jr., Twenty-sixth Division. Nogent-le-Roi, Marne, France, November 26, 1918. Courtesy of the National Archives.

(*Above*) President Wilson's Christmas dinner with the American Expeditionary Forces (AEF) at Montigney-le-Roi, France. *Left to right:* Rear Admiral Cary T. Grayson (American navy), Wilson aide and physician; Major General McAndrews, chief of staff, AEF; Mrs. Woodrow Wilson; General Pershing; President Woodrow Wilson; General Hale (American army), commanding the Twenty-sixth Division; Ambassador Jusserand, the French ambassador to Washington; General Hunter Liggett (American army); Madame Jusserand; General Summerall (American army), commanding the Fifth Army Corps; Miss E. Benham, secretary to Mrs. Wilson; General Leorat (French army). December 26, 1918. Courtesy of the National Archives. (*Below*) Summerall and Major General George Duncan at the Eighty-second Division review, Prauthey, France, February 1, 1919. Courtesy of the National Archives.

(*Right*) Summerall at the ceremony awarding him the Distinguished Service Medal. Biesles, Haute Marne, France, April 12, 1919. Courtesy of the National Archives.

(*Below*) Council of Four peace conference. *Left to right:* Signor Orlando, the Italian prime minister; Mr. Lloyd George; M. Clemenceau; and President Woodrow Wilson. The Paris home of President Wilson, 11 Place des États-Unis, Paris, May 27, 1919. Courtesy of the National Archives.

(*Above*) General John J. Pershing and officers of his staff aboard the SS *Leviathan*, en route to the United States from France. September 6, 1919. *Left to right:* Major Lloyd Griscom; Colonel Adalbert de Chambrun of the French army; Brigadier General Robert C. Davis; Lieutenant Colonel Albert S. Kuegle; Major General Charles P. Summerall; Captain James C. Hughes; Major General John L. Hines; Colonel George C. Marshall; General John J. Pershing; Captain George E. Adamson; Major General Fox Connor; Colonel John Quekemeyer; Major General Andre Brewster; Lieutenant Ralph A. Curtin; Major General Walter A. Bethel; Colonel Aristides Moreno. Courtesy of the National Archives. (*Below*) USS *Leviathan*. March 3, 1919. Courtesy of the National Archives.

(*Above*) Major General Charles P. Summerall addressing the troops after a parade in honor of the First Division, 1920. Courtesy of the National Archives. (*Below*) Summerall and Brigadier General William Mitchell during Mitchell's December 1923 inspection visit to the Hawaiian Department. Courtesy of Charles P. Summerall papers, box 35, Library of Congress.

(*Above*) Dedication of the First Division Memorial, with the State, War, and Navy Building (now the Eisenhower Executive Office Building) in the background, 1924. Courtesy of the National Archives. (*Below*) Summerall as Second Corps Area commander meets with New York City mayor Hylan, January 17, 1925. Courtesy of the National Archives.

9,792

(Above) Major General Charles P. Summerall, Secretary of War Dwight F. Davis, Major General J. L. Hines. November 20, 1926. Courtesy of the National Archives.

(Right) Major George C. Beach Jr., medical corps, receiving the Distinguished Service Medal from Secretary of War Patrick J. Hurley, for services in caring for patients at Camp Greene, NC, during the influenza epidemic of 1918. State, War & Navy Bldg. State, War, and Navy Building (now the Eisenhower Executive Office Building), January 23, 1930. Major General Charles P. Summerall, witness. Courtesy of the National Archives.

(*Right*) Major General Charles P. Summerall, Major General Merritt W. Ireland, surgeon general of the U.S. Army; and Major Fletcher, First Medical Regiment, First Division, inspecting new field medical equipment, Washington, DC, 1927. Courtesy of the National Archives.

(*Below*) Summerall laying a wreath at the First Division Memorial, May 30, 1927. Courtesy of the National Archives.

General Charles P. Summerall, chief of staff, the wartime commander of the First Division, places a wreath on the Tomb of the Unknown Soldier on behalf of veterans of the division during a reunion at Washington, DC, October 23–25, 1930. Courtesy of the National Archives.

Formal portrait of Summerall on his retirement as chief of staff, 1930, inscribed to William M. Steamer, his wartime orderly. Courtesy of the First Division Museum at Cantigny.

The Eustis, FL, home purchased and briefly occupied by the Summeralls subsequent to army retirement and prior to the move to Charleston, SC. Courtesy of the First Division Museum at Cantigny.

General and Mrs. Summerall early in retirement, ca. 1930s. Courtesy of the First Division Museum at Cantigny.

Laura Mordecai Summerall. Courtesy of the First Division Museum at Cantigny.

the Fifth Corps to resume the attack on October 9. At this time, a letter was written by the chief of staff of the First Army[3] to all division commanders urging them to visit their lines, maintain close touch with the men, and influence the action by their leadership. At the bottom of the letter to me he wrote: "This does not mean you." The 361st Infantry of the Ninety-first Division, much weakened by combat, was attached to the First Division and supported the right flank at Hill 269.

The plan for October 9 was for the right brigade to attack obliquely and occupy the extended front to Hill 269. The First Battalion, Eighteenth Infantry, which had been held in division reserve, was to replace it and assault the formidable Hill 272. A provisional squadron of cavalry with the division was used for liaison duty, although the terrain made mounted combat impossible.

In order to secure a maximum density of artillery fire, the attack was to be made in three echelons, with all the artillery firing a barrage in front of each echelon. Heavy concentrations of artillery were to precede the advance. On the morning of October 9, preceded by a heavy artillery fire along the entire front, all of the artillery dropped a barrage in front of the right brigade at 8:30 A.M. and led its assault on its objective. After heavy fighting, the enemy was driven back and the objective occupied beyond Hill 263. The artillery then delivered a heavy fire in front of the Sixteenth Infantry and with a dense barrage led it over Hill 272. This was one of the hardest-fought actions of the war. Many prisoners and machine guns were captured. In turn, the artillery then concentrated in front of the left brigade and led it to its objective. In the meantime, the Germans viciously attacked Hill 269, but the First Engineer Regiment held. During the afternoon, the 361st Infantry occupied the eastern slope of Hill 269, supported by a battalion of the 362nd Infantry. At the end of the day, the positions occupied extended just north of Fléville, along the north edge of the Côte de Maldah, to Hill 269. Enemy artillery shelled our area intensely night and day.

During the tenth, orders required exploitation only. The enemy had retired to such an extent that our lines advanced from one to two kilometers without serious resistance. During the day, I overheard a telephone conversation between the chief of staff of the division, Colonel Greely,[4] and the chief of staff of the First Army.[5] The former said: "The First Division never asks to be relieved. The high command knows the condition of the division as well as we do.

When the high command wants to relieve it, we will go. Until then, we will continue the battle." This reflected the spirit of the entire division.

After Soissons, I enunciated the following principles, which the division followed: The First Division never asks to be relieved; it is never tired, for it can take one more step; it is never held up by machine-gun fire; it never criticizes its neighbors. The enemy came to respect our prowess and bravery, as the following intelligence report, published October 10, 1918, indicated:[6]

Today a captured colonel of the German Army arrived at our Division cage. He was cold, hungry and broken in spirit. After four years of severe fighting and constant service in his army, he was taken prisoner by the victorious troops of the 1st Division. The following is the substance of his remarks:

"Yesterday I received orders to hold the ground at all costs. The American barrage advanced towards my position and the work of your artillery was marvelous. The barrage was so dense that it was impossible for us to move out of our dugouts. Following this barrage closely were the troops of the 1st Division. I saw them forge ahead and I knew that all was lost. All night I remained in my dugout, hoping vainly that something would happen that would permit me to rejoin my army. This morning your troops found me, and here I am, after four years, a prisoner.

"Yesterday I knew that the 1st Division was opposite us and I knew that we would have to put up the hardest fight of the war. The 1st Division is wonderful and the German Army knows it. We did not believe that within five years the Americans could develop such as the 1st Division. The work of its infantry and artillery is worthy of the best armies in the world."

The above tribute to the 1st Division comes from one of Germany's seasoned field officers. It is with great pleasure that we learn that even our enemies recognize the courage, valor and efficiency of our troops. The work done by the 1st Division during the past few days will go down in history as one of those memorable events, which will live in the hearts of American people for generations to come.

Every member of this command well deserves the enthusiastic congratulations from and the high respect in which it is held by our comrades in arms and by the entire American nation.

The above will be published to every member of this command.

The powerful effect of the artillery was gained by having all of the guns in the division fire over a segment of the front as an echelon of the infantry advanced. We had not nearly enough guns for the division to attack at one time. The recognition by the German colonel of the coordination and teamwork of the infantry and artillery was a realization of my great purpose since I entered the service.

On October 11, patrols located the enemy's new line, which was strongly fortified. Initially, we received orders to hold our lines. But, later that day, the division was ordered relieved by the Forty-second Division—on the night of October 11–12. We left details to bury the dead. At the same time, I was promoted to the command of the Fifth Corps. I ordered the First Field Artillery Brigade to remain in action, and it was not relieved. Some guns had started out and were ordered back to their positions. I needed them for the early assault. This was a great test of fortitude, but the men accepted it with good faith and courage.

From the time it had entered the Meuse-Argonne attack, the division had advanced seven kilometers and defeated elements of eight German divisions, including some of their best. Throughout the advance, both flanks were exposed to enemy fire. It captured 26 officers, 455 noncommissioned officers, 924 men, 13 field guns and mortars, and large numbers of machine guns and rifles. Many enemy dead were on the field. Our losses were 68 officers and 1,526 men killed, 126 officers and 5,706 men wounded, 59 men missing, and 33 men prisoners. Frederick Palmer called it *Our Greatest Battle.*[7]

It was indeed the decisive battle that ended the war, and, as at Soissons, the First Division bore the brunt and made victory possible.

General Pershing, in his famous General Orders No. 201, cited the division with the immortal words: "The Commander in Chief has noted in this division a special pride of service and a high state of morale never broken by hardship nor battle." This is the only AEF order during the war that cited a single division. No troops ever deserved more the gratitude and the reverence of their country.

In my visits to the line, it was heartbreaking to see the large numbers of dead on the field and around the dressing stations. In returning to my headquarters one night, I tried to avoid the bodies at one point in the trail, and I became lost in the woods. The firing was heavy, and the directions of the sounds were confusing. Presently, I heard voices speaking German, and I thought that I must have wandered into the German lines. I had taken only a runner as a guide, and he was equally confused. Then I heard a command in English and realized that it was a group of prisoners being sent to the rear under sentinels. I followed them and found the road.

It was necessary, after my appointment on October 11, for me to proceed at once to the headquarters of the Fifth Corps, and I could not see the men when they came out of the line. I published a farewell order to the First Division that tried to tell the officers and men of my pride and gratitude. No words could ever express the courage and sacrifice of that historic and victorious field.

It was with a heavy heart that I left the command post at midnight and started to the headquarters of the Fifth Corps. The road was congested with miles of stalled ambulances, and I knew the delay in reaching the field hospitals would cost many lives. Next to winning the battle, our chief efforts were to bring the wounded to the dressing stations and then move them to the field hospitals. Often we could not locate the groups of wounded who had received only first aid, and no doubt many died from delay in evacuating them.

I reached the command post of the Fifth Corps with my two aides, orderly, and driver about two o'clock on the morning of October 12. The corps commander whom I was relieving had been one of my instructors in drawing when I was a cadet at West Point.[8] I had always liked him. When he had visited my command post during the battle, he said that he never interfered with his division commanders. From what I afterward learned, he had told the commander in chief, his classmate at West Point,[9] that his corps could not accomplish all that was expected, and orders were given to relieve him. When I arrived, he was alone and walking back and forth in the small dugout. He shook hands with me and said: "Good-bye, I hope you will do better than I have done." He entered his waiting car, and I never saw him again. He was a crushed and brokenhearted man. I knew none of the corps staff consisting of about eighty officers.

The chief of staff[10] came in and said that he was leaving for an army conference of chiefs of staff. A large map of the corps front was on the wall, and I proceeded to study it. I called the chief of staff at the First Army, General Drum, and told him that the First Division should have a GHQ citation, not only as a recognition of what the men had done, but also to hearten them in preparing to reenter the line very soon. He told me to write what I thought should be said. I did so and sent it to him. The citation was a copy of my draft with the addition of the paragraph:

> The commander-in-chief has noted in this division a special pride of service and a high state of morale never broken by hardship nor battle.

The effect on the division was all that I had anticipated, and it is engraved on the hearts of the veterans while they live.

The corps chief of staff returned with orders to renew the attack on the morning of October 14. The corps front was held by the Forty-second and Thirty-second Divisions. The Forty-second was my old division (I had commanded the artillery brigade), and it occupied the First Division sector, with which I was familiar. I, therefore, went immediately after breakfast, where I met the general staff, to visit or inspect the Thirty-second Division. I called on the division commander,[11] who was a first classman when I was a plebe, and who was a lieutenant in the Fifth Artillery with me at the Presidio of San Francisco. I told him that I would like to go without him to see the troops, to which he assented. I had learned at corps headquarters that each brigade had a regiment of the other brigade in its sector, thus mixing the commands. I found both the brigade commanders at the command post of the first one that I visited. One had been a cadet ahead of me[12] for one year at West Point, and the other[13] I had known slightly in the Philippines. I told them that during the night they must recover their own regiments and unite their brigades. I also told them to organize their lines in depth with only a thin screen in the front line. One of them did not understand at first, but, when I fully explained, both eagerly adopted my suggestions. I then went without them to see the men and say a few words to each echelon. One regiment was commanded by a classmate[14] at West Point, and he was badly gassed. Indeed, the entire area was heavy with gas, and many men were affected.

In accordance with army orders, the Fifth Corps attacked at 8:30

A.M., October 14, preceded by a heavy artillery preparation, which began at 5:30. The reorganized Thirty-second Division fought gallantly and captured the town of Romagne and the Bois de Bantheville. The fighting was hard and lasted all day. The possession of the Bois de Bantheville was necessary as a jumping-off place for the final drive.

The right brigade of the Forty-second Division under Brigadier General Douglas MacArthur gained its objective, which was near the Côte de Châtillon. The left brigade gained very little ground and was stopped before Landres-et-St. George, even though this attack was supported by the artillery of the First, Forty-second, and Thirty-second Divisions.[15]

Early on October 15, the Forty-second Division resumed the attack. The right brigade occupied the foot of the Côte de Châtillon and partly surrounded La Tuilerie Farm. Again, the left brigade failed to advance. As soon as the lines were stabilized for the day, I went to see the troops of the Forty-second Division. My worst apprehensions were realized. At division headquarters, I found that the division commander[16] had never left his command post, which was far to the rear, and knew almost nothing of the situation at the front. The commander of the left brigade[17] was confused and completely unstrung. He knew nothing of why the brigade had not advanced and had never left his command post. I saw that he was in no condition to remain in command and decided to relieve him. On arriving at the command post of the nearest regiment, I found the colonel[18] starting to brigade headquarters. He knew nothing of why his regiment had not attacked and had never left his command post. He said that nearly all of his regiment had been killed or wounded and that many bodies were caught in the enemy's wire. I asked him how he knew this, and he said that some young staff officers had told him. He was mentally defeated and in no condition to command. Of course, the reports were not true, and there had been few casualties. The lieutenant colonel[19] was with the front line, I was told, and had been wounded. It was necessary to relieve the colonel,[20] thus incurring the hostility of the regiment, but he could not be trusted with men's lives. I decided to have the regimental machine-gun officer,[21] who was a colonel, take command, not knowing how seriously the lieutenant colonel was wounded. I found the other regimental commander[22] far back from his regiment. He had never gone to his lines and did not know why the regiment had not attacked. He seemed well poised and to realize he had failed in his leadership. I explained what I expected of him and told him to move his command post well to the front and to visit his line at once. He assured me that he would do so. In the part of the line I saw the troops were in good condition. It was evident that no real effort had been made to convince the troops that they were expected to attack. The men were as good as any in the army but needed leadership.[23]

I returned to division headquarters and told the division commander[24] what I had seen. I directed him to move his command post forward and to visit his commanders and lines at once and to report to me by telephone when he returned. I told him that if he did not do this, he knew what I would be compelled to do. He seemed responsive, but, although I had known him pleasantly for years, I realized that I had made an enemy of him. This I found later. I should have relieved him. I also told him to relieve the brigade commander,[25] who had become unsuitable, and to assign a colonel of an artillery regiment[26] whom I had long known to his place. Also, I told him to relieve the colonel of the infantry regiment[27] and replace him by the division machine-gun officer.[28] The brigade commander whom I relieved was a close friend and classmate of the army commander,[29] whose hostility I thereby incurred and who tried to hurt me later. It distressed me greatly to be compelled to take these measures with my old division, but I could not entrust the lives of men and the success of the battle to officers who had lost all ability to command. The attack was resumed on October 16 by the Forty-second Division, and some ground was gained. Orders then required that only exploitation should take place in preparation for another phase of the battle.

On October 18, the Eighty-ninth Division, which had not been in line, relieved the Thirty-second Division. I had the officers assembled and explained their duties and the necessity for their leadership in combat.[30] My theme was that at all times every man must be under the eye of a leader in the appropriate grade and that men were held up by fire only when they wanted to be held up. As I talked, a drenching rain poured on us, but we thought little of it.

The following account is taken from the History of the Eighty-ninth Division:

> The Corps Commander, Major General Charles P. Summerall, made daily visits to the Division, often assembling officers and men and speaking directly to them.

The talk of General Summerall to the assembled officers was in tone grim, blunt and somber the talk of a fighting man to fighting men. In substance, he spoke as follows:

"When a division enters the line in attack, it is given an objective to take. That objective must be reached. There is no excuse for failure. Either you take your objective or you do not take it. Casualties among officers will be heavy, as well as among men, although probably eighty per cent of the wounded will come back. Officers must keep well to the front and when anything goes wrong, it is the duty of the next commander to go up and see what is the matter. The toll of casualties of the senior officers will of necessity be increased by this practice but the results are more than commensurate with the costs. Control is vital. Divisions have been frittered away by straggling or the pernicious practice of sending details to the rear. In this corps, it is the order that no riflemen are to be taken away from a company for any purpose. The best way to safeguard the wounded is to push ahead and defeat the enemy. Pitiful examples have recurred in the present offensive wherein units have allowed their strength to be weakened by details for carrying wounded and, in the face of a counterattack, have been driven back leaving their wounded to die. To halt plays the enemy's game, since he is fighting a defensive action with machine guns and artillery. To halt means losses. But if you push on, the losses will not be much greater and you will have gained something. No officer should ever say that he is tired or allow his men to say it. No man is ever so tired that he cannot take another step forward.

"Don't ask for relief. Those in higher command are constantly considering the matter of relief. It is expected that the full measure of the organization's strength will be demanded of it before it is pulled out. It must be so if we win. When you have reached the stage that the gains you are making do not justify the losses you are sustaining, you will be taken out. Don't worry about your flanks. Distribution in depth protects them. Troops must hold their ground. To fall back, allows the enemy to play his guns on you causing losses and those losses, with the ones you have sustained in the advance, will be in vain. But even a squad or a platoon, if it holds its ground, will enable the whole line to advance. In the last few days, a patrol of twenty men, by fighting and holding its ground on a hill, enabled a whole division to advance. The best way to take machine guns is to go and take 'em. Press forward. The finest tribute that can be paid to a division is: 'It takes its objectives.'"

General Summerall concluded his talk with the characteristic remark that he would get down to see us as often as he could. That he would try to see us if things went well. If things did not go well, we would certainly see him soon.[31]

Such appeals as these had a marked effect on the conduct of the division in battle. It would be hard to conceive of a better fighting spirit than was shown by all ranks. The lesson of control was well learned; the total of straggling for the entire division was only a fraction of 1 percent.

It might be said here that the conduct of the Eighty-ninth Division throughout the operation was beyond praise. Twice it was notified that it would be relieved but on the urgent request of the division commander[32] was allowed to continue the battle until the Armistice. I believe that knowing what to expect and what to do was responsible for its brilliant success. Again, I may emphasize that troops fail from a state of mind rather than from enemy resistance. During the entry into the lines that night, the division had some casualties from interdiction fire from long-range Austrian artillery.

I had no transportation to move the Thirty-second Division to the rear, and the men were not in condition to march far. Moreover, I had no other reserve, and, because the enemy was very active, it was necessary to keep the Thirty-second Division near enough to use in case we were attacked. I ordered them to an area out of range of guns but possibly exposed to some airplane bombing. Arrangements were made to assemble all units, many of which had become mixed, and have them supplied with hot food, deloused, and issued fresh clothing. Before the division could move, the division commander[33] came to my dugout early in the morning when, fortunately, I was alone. He was greatly excited and at once began to rave and use the most violent language. He said that his division must be taken at least forty kilometers to the rear, where they could rest and be secure. He said many other things with profanity and gross defiance. I let him come to a stop and then replied as calmly and in as low a voice as possible to offset his shouting: "General, you and I have been friends for a great many years, and we are going to continue to be friends. You know that I can understand temperate language as well as violent language, and you know that you cannot talk to me in that manner." His whole manner changed, and he looked crushed. He said: "I beg your pardon. I should not have talked that way." Then I explained why I could not move the division far and that everything would be done for the comfort of the men. He went away subdued and calm, and to the end of his life he always showed me the greatest respect and friendship. Like the two colonels at Soissons, I could have relieved him and ruined his career, but because I recognized a condition of neurosis from overstrain and was patient and calm, he was saved to recover and be a good commander and officer for several years.

Early in the morning after the division was relieved, I visited all of the troops and told each battery and company of their fine conduct in line, praised their success, explained to them that I could not transport or march them further to the rear but that they would be well fed, rested, and deloused, and that everything possible would be done to prepare them to reenter the line as soon as they were needed. Their looks reassured me, and I felt that the officers and men understood and would accept the situation in a proper spirit. I had reason to know twenty-five years later that the effect was all that I could have desired.

I visited the Eighty-ninth Division the day after it entered the line. I asked the division commander to let one of his aides guide me. He was a handsome recent graduate of West Point.[34] On reaching the first reserve battalion, he spoke to a sergeant so roughly as to humiliate him. I told the young officer that I could not proceed and would return to the division command post. His manner would have antagonized the troops and spoiled my visit. I went next day with another guide and talked to all of the men from the reserve battalions to the front line. Thus, they knew me and understood my attitude toward them. The first officer became a major general in World War II. I did not let him suspect my reason for discontinuing my visit to the lines with him. I would not humiliate him.

From that time until November 1, the operations were in preparation for the great assault of the entire Western Front, first planned for October 28 and then changed to November 1. There was a strong attack by the Eighty-ninth Division to gain a jumping-off line at the edge of the Bois de Bantheville, and the Forty-second Division advanced to near Landres-et-St. Georges and St. Georges. With the reorganization, the Forty-second Division functioned ably, though the division commander[35] was weak.

Conferences were held at corps headquarters with division and brigade commanders and staffs at which I explained the entire plan of attack. General Dickman intruded at once and showed his ignorance by saying he would use his artillery for concentrations. I invited him to lunch, and he talked vulgarly at the table. On November 1, as I predicted, his corps did not advance. He became bitterly hostile and tried to hurt me when the First Division obeyed the orders of GHQ and the First Army to capture Sedan. His use of artillery on November 1 could not have suited the Germans better if they had ordered it. The Second Division relieved the Forty-second Division on the night of October 30–31.[36]

On October 30, I visited all of the regiments and explained the nature of the attack and told them that it would end the war if successful. I warned them against overconfidence and told them that while we knew they were good troops, other good troops had not been successful in their sector. The Ninth Infantry of this division had been in China with me, and I had known the Twenty-third Infantry in Manila. I also reminded the marines that I had served with them in the Philippines and China. I knew the division and brigade commanders. On reporting to me, the division commander stated that the Second Division wanted to be relieved as soon as it took its objective. I told him that I could not discuss relief. This was an omen of what happened later. The division had been told by the chief of staff of the First Army[37] that the Fifth Corps would lead the assault and drive a deep wedge and that the Second Division had been given the part of honor. The morale of the Second and Eighty-ninth Divisions was very high. The First Division had received ten thousand replacements and had trained intensively since it came out of the line on October 12. It was assigned to the Fifth Corps as a reserve. I talked to men in all of the regiments, praising their accomplishments, and telling them that in the approaching battle we would end the war. I said that if any of the attacking troops faltered, or if the line was extended, I would at once put in the First Division. The division was, on my request, commanded by Brigadier General Frank Parker,[38] who was the ablest combat leader in the war. I talked to nearly a hundred thousand men in the two days before the attack. My voice was very hoarse, but I was certain of success.

On reaching my command post late on October 31, I found General Pershing and the French general Maistre,[39] who commanded the group of armies adjoining us. When I spoke, General Pershing asked if I were sick. I told him that my voice was hoarse from talking to the troops. He asked what I thought of the prospects of the attack. I told him that I would give him every foot of ground on schedule. The French general said that the American losses were very heavy. I told him that if the French fought as hard as the Americans, their losses would be heavy too.

The success of the assault depended on the covering artillery fire to protect the infantry from the German machine guns and artillery at the beginning and as the line advanced. I planned to use every weapon to its maximum power. I ordered all machine guns to fire six hundred shots a minute and all 75-millimeter guns to fire four shots a minute. For the attack, we had 272 75-millimeter guns, 180 155-millimeter howitzers, 124 155-millimeter guns, 24 8-inch howitzers, and 8 6-inch mortars, or a total of 608 guns. The artillery plan included fire for destruction from all the guns for two hours before the assault against the enemy's lines, artillery batteries, and rear installations. Two 155-millimeter guns were assigned to every enemy battery and fired gas and shells. The barrage plan was so constructed that throughout the advance, the entire corps front of more than eight kilometers would be covered by a sheet of shell, shrapnel, and bullets to a depth of twelve hundred meters. To this fire was added the machine guns of the Second, Eighty-ninth, and Forty-second Divisions. During the preliminary bombardment, every known enemy battery was to be silenced and the infantry pinned to their trenches. All known machine-gun positions were attacked by at least one howitzer. In the initial barrage, there was a line of bursting shell 150 yards in front of our infantry. Two hundred yards in front of the shell line was a line of shrapnel bursts. One hundred and fifty yards beyond the shrapnel was a machine-gun barrage of great density. Three hundred and fifty yards in front of the shrapnel was a line of bursts from 155-millimeter howitzers. Beyond this were concentrations on occupied points by 155-millimeter guns and 8-inch howitzers. The trench mortars covered the enemy trenches. Smoke shells in the initial bombardment and during the advance concealed our infantry. As the troops advanced, the artillery moved to forward positions for close support. The fire for destruction began at 3:30 A.M., November 1. At 5:30 A.M., the barrage stood for five minutes in front of the infantry line and then moved at varying rates according to the difficulties of the ground. This became known as "the Summerall Barrage." It has probably never been equaled.[40]

The effect was all we could want. All partial objectives, where troops halted for rest and the barrage stood in front of them, and all final objectives were taken on schedule. When the Second Division reached its objective for the day in front of the Heights of Barricourt, the division commander[41] asked what he should do. I told him to continue during the night through the Barricourt woods and gain the further edge before morning. This greatly surprised the Germans, and, besides saving hard fighting, many prisoners were taken. The Eighty-ninth Division advanced abreast of the Second Division most gallantly. The First Corps on our left did not gain any ground. Only a few individuals joined our left flank and followed. The enemy, being entirely outflanked by the Second Division, retired in front of the First Corps, and the next morning the corps boasted of sending men in trucks to regain contact. To me, the inaction of the First Corps was unpardonable, but it was glossed over by friends at GHQ.

Our losses were very light, while those of the enemy were heavy in prisoners and killed. So effective was our fire that we found enemy machine guns with belts inserted but not a cartridge fired. There also were field guns with muzzle covers on that had not fired a shot. Whole teams and drivers of artillery lay dead where they tried to withdraw. This was the only place where the artillery approximated the strength that I had advocated at Paris[42] for attack, and the saving of the infantry was proved as well as the overwhelming of the enemy.

As soon as the objectives were reached, I visited the lines. En route I stopped, as always, at the field hospital and talked to the wounded. I knelt by one stretcher where a man lay and said: "I am the corps commander. I have come to thank you for what you did today and to tell you that all of our objectives have been taken." He opened his eyes, and his face lighted with recognition, and he said: "Oh, yes. I met you yesterday." I was thrilled by the realization that he was only one of the thousands to whom I had spoken, but I had made him, and no doubt many others, feel that I was so personal that they had met me. The success of the assault was simply an application of the age-old principle of fire and movement.

I wrote letters of commendation to the divisions, encouraging them to continue the advance. It was probably here that the bitter hostility of General Dickman, whose corps did not advance, was aroused by jealousy. As General Horace Porter[43] said at the reinterment of the remains of John Paul Jones: "Success is like the sunshine; it brings the vipers out."

The advance continued against heavy resistance. The Second Division captured Beaumont on November 5 with heavy losses. A wide gap soon occurred between the left of the Fifth Corps and the right of the First Corps. This was filled by the Twenty-sixth Infantry of the First Division, which came into line at 5:00 A.M., November 5. The First Division relieved the Eightieth Division of the First Corps at 5:30 A.M., November 6, and immediately attacked in the direction of Mouzon on the Meuse River. The fighting was severe, but all objectives were taken, with the right of the division reaching the Meuse River. On the morning of November 6, the Fifth Corps received the following:[44]

Memorandum for Commanding Generals, 1st Corps, 5th Corps.

Subject: Message from the Commander in Chief.

1. General Pershing desires that the honor of entering Sedan should fall to the First American Army. He has every confidence that the troops of the Ist Corps, assisted on their right by the 5th Corps, will enable him to realize the desire.

2. In transmitting the foregoing message, your attention is called to the favorable opportunity now existing for pressing our advantage throughout the night. Boundaries will not be considered binding.

By command of Lieutenant General Hunter Liggett
H. A. Drum.
Chief of Staff

I immediately took copies of the order to the three division commanders. I reached the First Division headquarters shortly after noon. The division was still advancing, and all of its front was not known, but elements were expected to reach the Meuse River. To me, the manifest thing was for the division to cross the Meuse wherever the river was reached that night. It did not occur to me that the division commander would do otherwise. However, nothing was known of the conditions for crossing, and no reconnaissance had been made. I could not give detailed orders for the execution of the memorandum but told the division commander[45] that I expected him to be in Sedan the next morning. He replied: "I understand, Sir. I will now give my orders." He and his French liaison officer began studying the map. There was nothing more that I could do, and it was necessary for me to proceed to the other two divisions. I was amazed to learn on the following morning that the division commander had withdrawn his troops from the Meuse and had moved the division down the area between the Meuse and the Bar Rivers. In doing so, he had entered the territory occupied by the Forty-second Division of the First Corps. When the order reached the troops to move, one captain refused, saying that the First Division never gave up any ground. When I reached the division on the morning of the seventh, I found the right brigade engaged in line with troops of the Forty-second Division and the First Division commander with a few of his staff near the front line in a command post. I at once asked him to break off the action, move all of the First Division to the right of the First Corps, and cross the Meuse that night along his immediate front. On reaching the rear command post of the division, I received a telegram from the First Army to withdraw the division to the rear at once. In this operation, the division had lost 2 officers and 127 men killed, 5 officers and 218 men wounded, and 2 missing.

On my way back, I met the commanding general of the Forty-second Division[46] and his staff, several miles in rear of the line, driving rapidly. I tried to stop him, but he continued. Afterward, I had reason to believe from his reports that he was smarting with resentment because I had ordered him to see his troops and carry out my orders to attack after failure on October 14 and 16. A few miles further back, I met the commanding general[47] of the Sixty-seventh Field Artillery Brigade of the Forty-second Division and his staff. I signaled him to stop, which he did. He said that the division headquarters was moving to Maison Blanc, some miles in rear of the line, that he had "shot hell" out of the Germans, and that all was going well. There was no confusion in the lines where I had been, and the artillery brigade commander who was with division headquarters had no idea of any confusion. It is certain that the commander of the Forty-second Division had never been near the lines. From this, there grew the most bitter controversy of the war over the execution of the order to capture Sedan. The action of the First Division commander in moving between the Meuse and the Bar was strictly in accordance with the first order received dated November 5. His entering the zone of action of the First Corps was strictly in accordance with the second memorandum, which said that boundaries would not be considered binding. The Forty-second Division had not advanced, so there was no confusion or interference. Had he tried to cross the Meuse where he was on the evening of the sixth, he might have been censured for not obeying the orders in case of failure.

On reaching my headquarters, I found violent criticism from General Liggett, commander of the First Army, originating in complaints from General Dickman, who had never been within many miles of the troops, and General Menoher, who, to my knowledge, was far behind the lines. When shown the order for the advance, General Liggett said that he had never heard of it. Immediately after the Armistice on November 11, I visited General Dickman's headquarters many miles behind the lines. He was abusive and said that I had played hell with his corps. The only contact the First Corps had with the First Division was to give the troops some rations far back of the lines and is set forth in the history of the Forty-second Division, called *Americans All.*[48] It is the only incident in history where troops have advanced strictly in obedience to orders, defeated the enemy, and been criticized by those who gave the orders.[49]

Orders were then received for the Fifth Corps to cross the Meuse and advance, but the army had no boats. I ordered the Eighty-ninth Division to build some catamarans from captured German bridge equipment, and I had floating bridges made for the Second Division. I selected 9:00 P.M., November 10, just as the moon went down, for the crossing. Conferences were held, and all details were understood. On November 9, I received the following order from Marshal Foch through the First Army: "The enemy, disorganized by our repeated attacks, retreats along the entire front. It is important to coordinate and expedite our movements. I appeal to the energy and initiative of the commanders-in-chief and of their armies to make decisive the results obtained."

At 9:30 P.M., November 10, the artillery of the Fifth Corps dropped a heavy barrage on the further bank of the Meuse River and attacked the German batteries. The catamaran took a company at a time across the Meuse, and the troops seized the enemy's position. The Second Division crossed as rapidly as possible on the floating bridges, which were carried on the shoulders of men and shoved across the river. By morning, eleven battalions had crossed and had advanced up to a mile from the river.

At 8:30 A.M., November 11, I received an order to cease the advance at 11:00 A.M. as an Armistice had been declared. I was greatly amazed and regarded it as only a delay gained by the Germans. The order was sent to the troops, but it did not reach the leading troops of the Eighty-ninth Division till 12:00 noon. I at once visited the troops. I had ordered a bridge built at once by the Eighty-ninth Division engineers. When I reached the division, I found that nothing had been done. I sent for the commanding officer of the engineer regiment[50] and asked why he had not obeyed the order. He replied that it was against international law. I saw that he was not in his right mind and ordered him to the rear. I then ordered the engineers of the Second Division to build it. By night, we were able to send trucks across with food and ammunition. I went to the front lines and talked to the men. Their spirits were very high. Our losses had been light and no doubt much less than we would have suffered from enemy artillery fire in the positions that we left. I commended the men for what they had done and told them that we must be ready to resume the attack. That night, fires were built, and the men slept.

I learned later that the brigade of marines in the Second Division resented the order to cross the Meuse River, as the Armistice came immediately afterward, and blamed me for their losses, which they said were unnecessary. Of course, I knew nothing of an armistice and was obeying the most urgent orders from the high command. The selection of the brigade of marines to lead the crossing of the Second Division was done by the Second Division commander, General Lejeune, a marine, without any reference to me. No doubt, he wanted to gain credit for the marines. There was no complaint from the army brigade in the Second Division. From my experience with the marines, they should never be employed with the army or under army officers. They fight no better than the army, and they complain, seek quick relief, and try to gain publicity.

During the battle, visiting officers stated that the thought existed in the army that a Third Army would be formed and that I would command it. A report was also made to me that Secretary of War Baker had stated that if General Pershing became disabled, I would take his place. I had no ambition and was always entirely satisfied with any command that I had. However, the part taken by the Fifth Corps led to the belief that we ought to lead the advance into Germany. We were much surprised, therefore, when our divisions were assigned to other corps and marched to Germany while our headquarters were ordered to Nogent-en-Bassigny. The sunshine of our success had brought the vipers out at GHQ. There at once came to my mind the fate of General Gouverneur K. Warren,[52] who was the brains and the ablest leader in the Army of the Potomac. When his Fifth Corps gained the lost victory at Five Forks and the war ended, he was relieved by Sheridan with Grant's authority because his successes had brought the vipers out. To further belittle us, we and our troops were spoken of with contempt at GHQ as only combat soldiers, belonging to an inferior category.

About November 14, the corps headquarters proceeded to Nogent-en-Bassigny, where the Twenty-sixth, Twenty-ninth, and Second Divisions were assigned to it. Of course, all of the troops not ordered to Germany were eager to go home. The problem was to retain morale during the long wait. A schedule of training was issued, including large-scale maneuvers, followed by critiques for the officers. Horse and transportation shows and division athletic and other contests were frequent. Here, as always, I spent the days visiting the divisions and encouraging them. The results were all that we could desire. There was never a thought of revolt. Discipline, dress, bearing, and soldierly conduct were excellent. Leaves were liberally given to men and officers.

When Christmas was approaching, I was notified that President Wilson would take Christmas dinner with a division of the Fifth Corps. I was thrilled at the possibilities. I imagined that he would be gracious to the troops and would say something to hearten and encourage the entire army. I selected the Twenty-sixth Division to be his host and suggested that the division be represented at the dinner by one hundred enlisted men who had received the Distinguished Service Cross. I was informed that the president would not dine with enlisted men but wanted officers at the dinner. The division commander[53] invited me to attend. Mr. Wilson reviewed the division and went to Montigny-sur-Rois for the dinner at 12:00 noon. I was standing in the door to receive him, but he brushed past without speaking, though he had known me at West Point and in Washington. The mayor of the town then asked him to sign the Golden Book, and he declined. The mayor with great firmness told him that he could not visit his city without signing the Golden Book, and, with very bad grace, the president signed it. He remained at the table a very short time and left without saying anything to those present. It was clearly a case of a man drunk with power. He was in no way the man I had talked with several times at West Point and when he became president.

Orders from the army required that crosses be placed on the town halls in the cities occupied and illuminated at night during Christmas week. The one at Nogent-en-Bassigny could be seen for many miles and looked at night like a detached cross in the sky. The people flocked to the square at night and knelt before it. Toward the end of the week, a delegation begged me to leave it. They said that all crosses had been destroyed by the government in 1905, and they thought it a miracle for the Americans to come and restore it. Of course, we took it down at the end of the week.

At a field day of the Eighty-second Division, the program ended with a review for me. The troops were on a large plateau where it was said that Caesar had fought and narrowly escaped.[54] The ground was covered with snow. When the troops were formed for the review, the division commander[55] said that he had just received the report of a board of officers that showed that a Corporal York,[56] in the attack on the northern end of the Argonne Forest on October 8, had, single-handedly, killed or captured a German machine-gun battalion. I asked where Corporal York was. He said that he was the division standard bearer immediately in front of us. I told him to have the corporal ride to the front, which he did. I then cited him, and my voice carried to the entire division over the snow. The corporal appeared much surprised and rode back to his post. That night the division commander telephoned me for a copy of my citation, which I sent him. York was awarded the Medal of Honor, promoted to sergeant, and became one of the notable figures of the war.

In December, I was ordered to GHQ at Chaumont to receive the Legion of Honor. A number of general officers were decorated by Marshal Pétain, and I received the decoration of commander. General Pershing invited us to his chateau for lunch. As I approached the door, Marshal Pétain was leaving his car. I alighted from mine, and he came to meet me. He suggested that we take a walk in the park. After a few words, I asked him to tell me how he restored the morale of the French army when he took command after the disastrous French offensive in the spring of 1917. He showed eagerness to do so. In effect he said:

> I am a soldier and not a politician. Men respond to praise and not to blame. When I was a major, if my battalion did not execute my commands properly, I would say: "Men, that was my fault. I did not make my meaning clear to you." Then every man blamed himself and not me. I would repeat the order, and the men would do their best to execute it. When I took command of the army, I visited all of the troops down to the advanced observation posts. I praised the men and told them that they had done well. I found that only a few were allowed to go on leave, so I ordered that a large percentage of the army should be allowed to go home and should not be called back until their leaves expired. I found that they had no recreation, so I established canteens where the men could buy little luxuries cheaply and gather to see one another. I found that no recognition had been given to individuals or units for distinguished conduct. I had commanders recommend large numbers of men for decorations. I established the *fourragère* and awarded it to regiments and divisions who had been outstanding in their courage and combat. In a very little while it was a different army and was ready for combat.

At the lunch, I was surprised to find myself seated on General Pershing's right, as several of those present were my seniors. As soon as we began to eat, he said: "Summerall, tell me about that advance of the First Division at Sedan." I explained the orders that had been received and marked on the tablecloth the roads by which the troops had moved with the rivers and the principal towns. He followed closely and, when I had finished, showed every evidence of being entirely satisfied. I knew then that he had been told lies about confusion in the First Corps. Afterward, I marveled that he could receive so many false reports and treat them as such.[57]

Immediately after lunch, General Pershing went to his room. In the large reception room, I rejoined Marshal Pétain, and, in the course of our talk, I asked him if he intended to write a book on the war. He said that he did, and I asked him to send me a copy, which he said he would do. General Harbord,[58] who from May to July 1918 had commanded the Second Division, came to me and urged me to go to General Pershing and ask him to cite the Second Division. I was surprised at this because from June 1917 to May 1918 General Harbord had been General Pershing's first chief of staff and was very close to him. I asked General Pershing's aide to request General Pershing to see me, and I was conducted to his bedroom. He had removed his shoes and was sitting before a fire. I told him of the superior service of the Second Division and asked him to give it a GHQ citation. He seemed little interested and said practically nothing. I then withdrew and told General Harbord what I had done. Years later, I discovered that General Harbord was one of my chief enemies. The citation was not made.

When I returned to Nogent, a corps conference was in progress, and the officers of the Fifth Corps gave me a warm reception. At my billet, the owner called in great ceremony to congratulate me on being a commander of the Legion of Honor. Later, when I put the cross over the Hotel de Ville, he requested me to leave his house. I found that he was a leader of the socialists and was hostile to religion.

During the winter, I was called before a board of GHQ staff officers[59] to give my opinion on types of armament. I urged the adoption of the 105-millimeter howitzer as an all-purpose gun. I explained that every gun must be able to attack planes in order to fulfill the mission of artillery to destroy that element of the enemy that, at the moment, is most dangerous to its own infantry. I also pointed out the superiority of the 105-millimeter howitzer over the 75-millimeter gun. Of course, my ideas were not adopted.

I was made a brigadier general in the regular army February 16, 1919. At Nogent, I first saw the entertainers who had been visiting the rear areas during the war. A group, containing several young ladies, gave a musical entertainment one evening. When I entered the hall, I saw a number of small, ragged, dirty boys being driven out by the military police. I was told that these boys worked in the mines all day. Of course, I stopped their eviction and had them remain. A soldier heckled one of the young ladies, and I marveled at her tact in seeming to want to sing what he proposed. She evidently had tuberculosis, and I was told that she died shortly after the war.[60]

At the division maneuvers, I went among the men, and I often asked them who their officers were. Many did not know their platoon commanders, and the acquaintance diminished as the rank increased. Few knew the division commanders. I pointed out at a critique that an officer's personality, whatever it was, was his greatest asset and that officers should be seen and known by their men. One day, a division commander said that he had a joke on me. He had asked a soldier who I was, and the man replied: "Oh, he is the chief of staff."

During the winter, GHQ held a conference on the attack of November 1, 1918. The map showed the Fifth Corps bulge, with the First Corps far behind. The same situation existed on November 11. While still at Nogent, I also wrote a number of citations and letters of commendation to officers, men, and organizations. About the middle of March 1919, the Fifth Corps was demobilized. I took farewell of the able and loyal staff. In my remarks, I described our separation as sparks from the anvil. We went in many directions, and few ever met again.

My next assignment was to command the Ninth Corps at St. Mihiel. There, I found another new staff with few exceptions. The chief of staff[61] had been a second lieutenant with me in China. After a day or two, in which I visited the three divisions in the corps, I found that no papers or action had been placed before me. The chief of staff said that he handled all administration and that my predecessors had not been troubled with papers and decisions. I told him that all important matters of administration must be brought to my attention. We soon had a corps maneuver of the three divisions. When the officers assembled for the critique that followed, the chief of staff took charge and tried to show his superior military knowledge by referring to the campaigns of the ancients. I eventually injected myself into the proceedings and told him that thereafter I would conduct the critiques. It had not occurred to him that a corps commander was able to conduct a critique.

Early in March 1919, one of the divisions in the corps was ordered to the embarkation area. It was filled with resentment toward the division commander and the army inspectors who had reported adversely on the division on coming out of action, especially belittling it as a national guard division.[62] I invited the brigade and regimental commanders and the staff to dinner at my headquarters. After eating, I asked them to state their grievances, which they did most of the night. I then tried to show them that the combat record of the divisions should not be clouded on returning home by bitterness and revenge. I urged them to let the past rest with themselves and to emphasize to their people at home the pride and the service of the command. One officer said when I had finished that if they had had me as a commander, there would have been no grievances. In the division, one officer went to the U.S. Senate[63] for a number of years, one became secretary of war,[64] and one became president.[65] Many others later occupied prominent positions.

Later, GHQ ordered me to proceed to the embarkation area and try to improve the morale of another division, which was about to embark for the United States. I had known the senior officers when I was in the Militia Bureau before the war, and the division, or parts of it, had been under my command in France. The artillery had trained under me at Tobyhanna before the war. The brigade and regimental commanders and staff gave me a dinner. During most of the night, they poured out their grievances to me. Their worst complaint was the relief of a division commander,[66] to whom they were devoted, and the assignment to them at that time of a chief of staff[67] whom they hated. Again, I pointed out the fine record of the division and the harm that would be done to public opinion by returning home filled with bitterness instead of showing pride in their success. I told them that I would ask for a change in the chief of staff. This was done, and I had substituted an officer of the division whom I knew well and whom all liked.[68]

I was then detailed to accompany the Committee on Military Affairs of the House of Representatives to travel over the theater of operations and into the bridgehead in Germany. For several days, I explained the battles on each field. When at the Meuse-Argonne they asked me who was in command at a very hard-fought place, I told them that we had no heroes but that our real heroes were the men in the foxholes. During our trip along the Rhine, we visited General Mangin's headquarters of the French army. He told them prophetically: "Gentlemen, we have lost in the Council Chamber what we gained on the battlefield." Some of the friends I made on the committee were very helpful in my plans years later when I was chief of staff of the army and needed friends in Congress.

The Ninth Corps was demobilized, and I was ordered to command the Fourth Corps with headquarters at Cochem Castle in Germany, about forty miles up the Moselle River from Koblenz. The headquarters were in a twelfth-century castle on top of a high hill.

There were three divisions in the corps, and I spent much time visiting the troops.

Many visitors passed through, and we generally had a number to spend the night. One evening, I entered the great room where our guests assembled before dinner and found several ladies and officers. One lady was quite superior in appearance, and it was evident that I should escort her to dinner. I did not understand her name. When we were seated at the long table in the banquet hall, she said: "I have come to find the body of my son, and I have found it." She opened her bag and showed me the picture of a young officer whom I knew and who was killed at Soissons. We had found a grave with his name on the cross, but it contained the body of a German soldier. I supposed she had found that grave. She then explained that an officer had taken her to where he, a staff officer, had left the brigade command post to go to the front line. She then followed a route that led out of the First Division zone of action into the neighboring French sector and found a grave with a helmet bearing his name on the cross. The body was exhumed, and it was her son. I expressed my admiration of her fortitude, and she exclaimed: "How could I have done anything so wonderful as to raise such a son, and how could he have done anything so noble as to give his life for his Country!" She was the first mother I had seen who had lost a son, and I wondered if all mothers could be so resigned. She explained that in childhood and at Harvard, she had devoted most of her time to this only son and that they were so close she knew what he would do in going to the lines.[69]

I was selected as one of the American generals to witness the signing of the treaty of peace at Versailles, June 28, 1919. My seat was well to the front. The documents were open on tables. I went to them and saw the arrangements for the signatures. The hall was well filled with officers of the different Allies. A number of distinguished-looking ladies stood along the walls. At the appointed time, the members of the American Commission to Negotiate Peace entered and seated themselves at the tables containing the documents. Of course, I noted President Wilson, M. Clemenceau, Mr. Lloyd George, and others. Then the three German delegates marched down the long aisle with heads erect and defiant. They stood in front of the tables, and M. Clemenceau told them that they had been brought to bar for their crimes by the civilized world. The signing then took place. It was an impressive and historic occasion.

At this time, I learned that the American army would be sent home from Germany and that the First Division would be the last to leave. It was not in my corps, but I hastened to tell them of the honor. When I spoke to the Second Brigade, there were a few sounds of objection. I immediately told them that such conduct did not become the First Division, that this was a crowning honor, and that always they would proudly say that they were the first to go and the last to return. The rest of the division accepted the news stoically. They soon became proud of the assignment to leave last. I also arranged with the division to send a detail to erect the monuments containing the names of the dead on each battlefield in France. The money for them was contributed by the Salvation Army.

About this time, the First Division assembled at Montabaur, at its headquarters, and formed the Society of the First Division, AEF. The men gave me an ovation and elected me as president. At the dinner, I visited many tables and spoke a few words to each group. It was at this time that I was told by a Red Cross worker that the marines were very hostile to me because I ordered the Second Division to cross the Meuse River on the night of November 10. Nothing could have surprised me more. The orders had come from the First Army. The division commander selected the marine brigade to lead the crossing. There was no thought of an armistice. They lost fewer men in the crossing than they would have done in their exposed positions under heavy Germany artillery fire. They should have been proud of the honor of leading and of advancing at the hour of the Armistice. Years later, I was attacked by one of them in a newspaper article. I could only conclude that the marine may fight well but that he is undisciplined and lacks staying qualities. I would not want them in a command. It was the infantry brigade that gave stability to the Second Division.

Chapter 21

The Adriatic and Peace

When the Fourth Corps was demobilized, I was transferred to Koblenz. But I had been warned that I soon was to go to Warsaw as head of a mission to Poland. But then one afternoon I received a telegram to report to the American Commission to Negotiate Peace at the Quai d'Orsay[1] in Paris the next morning at nine o'clock. I drove nearly all night and reported as directed. The members of the commission were seated at a long table. Mr. Wilson gave no sign of recognition. M. Clemenceau said that the commission wanted me to proceed at once to Fiume as a member of an inter-Allied commission [the Inter-Allied Commission of Inquiry on Fiume] to adjust the conflict between the French and the Italian troops at that place. I was given no funds or assistance. I took an American and a French aide and my orderly, Sergeant William M. Steamer.[2]

The train was to leave late in the afternoon for Venice. During the morning, I received an invitation from a lady I did not know to a tea at the Ritz Hotel. As soon as I spoke to her, she introduced me to an attractive young Italian naval officer, Commander Bartolucci.[3] He at once asked if he could talk to me about Fiume. I told him that I did not wish to discuss it as I wanted to hear the case with an open mind. He then asked me to accept some documents, which I did. At the train station, Bartolucci appeared and said that his admiral in Paris told him to accompany me to Venice and be of any service to me. I found the French general Naulin, a member of the commission, on the train. Bartolucci asked me to dine with him on the train. I told him that I would do so if he also invited the French general, which he did. At Padua and other places, Italian officers came to the train to bring the compliments and offer the services of the local commanders.

As we approached Venice, Bartolucci said that a royal barge would meet me and take me to the hotel. I told him that I would accept only if he also took the French general and his staff. We went along the canals to the Grand Hotel. After dinner, a group of Italian

officers escorted and surrounded us to prevent us from being attacked in St. Mark's Square. The crowds were very hostile. Bartolucci said that an Italian destroyer would take me to Fiume. I told him
that he must also take the French general and his staff. No provision
had been made for us by the American navy. When we disembarked
at the dock in Fiume, I was conducted by Bartolucci to an Italian
car through thousands of Italian soldiers and civilians. As soon as I
was seated, a stick thrown through the front of the limousine struck
me in the chest. I pretended not to notice it. The crowd was very
hostile. When we reached the apartment where I was to be billeted,
an old lady who evidently occupied it protested vigorously against
having an American officer in her house. She was forced out by the
two Italian officers. There was practically no food in Fiume, and we
could obtain little to eat at dinner. An American cruiser, the USS
Pittsburgh, with the admiral in command,[4] and several small ships
were in port. The next morning, the lady who did not want me in
her house came early with a tray holding a delicious breakfast. I
asked Steamer what this meant. He said that he had bought some
food for her on the *Pittsburgh* and that she would now be friendly. I
made a courtesy call on the admiral on the *Pittsburgh.*

The commission met at in the city hall or capitol. It consisted of
Lieutenant General Robliant of the Italian army, Major General Naulin of the French army, an English major general, and me. The first
order of business was to elect a president of the commission. The
Frenchman and the Italian considered themselves disqualified. We
voted by ballot. The Englishman received two votes, and I received
two. Because I did not want to be president, I asked the Englishman
the date of his commission. It antedated mine. I told him that he was
senior and must be president. We then adopted French as the language of the commission. The Italian and the Frenchman spoke no
English, and the Englishman spoke no French. I could use French.
We studied our directions and adopted a course of procedure. On
going to the meeting, I was guarded by Italian soldiers. The crowds
on the street were very hostile, and the Italian guard on the door did
not salute me. I learned in Venice that the Italians were hostile to all
Americans because Mr. Wilson had arbitrarily drawn a line through
the middle of Istria and decided that the coastal half should belong
to Italy and the other half to Yugoslavia. The Italians claimed that
Mr. Wilson's division gave Fiume to Yugoslavia even though it was
entirely Italian in population and living. It was occupied by Ital-

ian troops with some French troops who had been attacked by the Italians. A street there had been named for Mr. Wilson immediately after the Armistice, and Fiume had the signs torn down.

Strategically, Fiume was the most important city on the Adriatic. The old Roman road and the railroad from the Danube reached the coast at Fiume. They were built in the great depression between the Jural and Dinaric Alps, which had been the route of barbaric invasions in ancient times. There was a harbor at Fiume formed by the quay, and on the other side was the harbor of Sušak, protected by another quay. Fiume and Sušak were close together, separated only by the Fiume River, a small stream. Sušak was all Yugoslav. The railroad continued along the Adriatic to Spalato. Since the reign of Maria Teresa of Austria, Fiume had been a *corpus separatum* in Austria. Since ancient times, Fiume had been largely populated by Italians, as was most of the Adriatic coast. There was bitter hatred between the native inhabitants of the interior and the Italians.

For six weeks we heard evidence in eight languages of the clash between the French and the Italians and the claims for the territory. The French wanted the right to exploit the timber in the interior and to use the port of Fiume for shipment. The Italians claimed it by right of occupancy and necessity. The Yugoslavs claimed it as a part of their territory under the Armistice. Through all ran the age-old hate and antagonisms.

I knew little of the situation and tried to treat it with an open mind. On the second day, Admiral Andrews invited me to lunch with him on the *Pittsburgh*. When I left the door of the palace, the Italian guard saluted me smartly. My car was entirely covered with flowers, and there was no guard to accompany me to the door. I asked my orderly, Sergeant Steamer, what had happened. He said that the Italians had found out that I was not hostile to them and that they wanted to show their appreciation. At the lunch, Admiral Andrews invited me to live on the *Pittsburgh* and to occupy the room of his executive officer, who was absent. My aides and orderly also lived aboard. Otherwise, I do not know where we would have gotten food. There was a nice little hotel at Abatzia, some miles from Fiume, where the members of the commission gave dinners to the others. The Englishman was very late to mine and did not show any regret. We had proceeded without him. I do not know how food was obtained by the hotel for these dinners, for there was almost no food in Fiume. I invited the Countess di Robilant and her daughter, who were visiting General di Robilant, to the dinner that I gave. After a few weeks, Commander Bartolucci asked if I would relieve him, as he was not well. He had attached himself to me, so, of course, I had no objection. When he left, I sent a box of cigars and some candy to the minister of marine in Rome with expressions of appreciation for his courtesies.

The weather was very hot, and the sessions of the commission were long. I studied the proceeding of the Allied admirals on the Adriatic during the war, the Italian-British pact [i.e., the London Pact][5] before Italy entered the war, which had been a secret document, and various other Adriatic documents. When we considered that all evidence was in, we decided to have each member of the commission prepare a set of findings and recommendations. The documents of the Englishman and the Frenchman were of little value. It was decided to take my paper and try to reach an agreement. My main recommendations were to give Fiume to Italy with the port of Fiume, to give Sušak to Yugoslavia with the port of Sušak, and to place goods from the Danube in bond at Fiume for shipment by water or rail. I had considerable trouble with the Englishman and the Frenchman, but they finally agreed. We parted very friendly. My dinner had much to do with it.

An American destroyer took us to Venice. I had the French general and his staff come with me. At Venice, I was able to shop with my aides, and I bought a shawl, a fan, some beads, and other lovely things for my dear wife. On reaching Paris, I was attached to the American Commission to Negotiate Peace and expected to be called on to tell the commission what had been done. After a few days, I went to each of the American members except Mr. Wilson and asked if I was to make a report. All said that they knew nothing of it, and nothing was ever done about it.[6] Within a short time, d'Annunzio[7] and the Italian troops forcibly took possession of Fiume and declared it Italian. In another year, a treaty between Italy and Yugoslavia embodied my recommendations. I might say that I never received any money for my expenses or those of my aides and orderly.

It consistently was stated in the hearings that the violent abuse of each other's countries by the French and Italian papers was the chief cause of hatred and friction between the two nations. I had met the daughter of M. Clemenceau, who was known as Madame Clemenceau. On reaching Paris, I asked her to dine with me at Ciro's. After a good meal, I told her that I supposed she wanted to know why

I had invited her. She said she could not wait any longer to hear. I then told her about the press and said that if the French would stop, the Italians would stop. I told her that the only man who could control the French press was her father and that she was the only person who could influence him. I asked her to do what she could to have him act. She enthusiastically agreed, and I believe that much good resulted. Commander Bartolucci called on me and said that he was very anxious to come to Washington as assistant naval attaché. I told him that I would do anything I could. He also brought a request from the Italian premier,[8] who was in Paris, to call on him. I did so, and he thanked me for my fairness to Italy. I told him that I would like him to send Bartolucci to Washington as assistant naval attaché, and he said that he would be glad to do so.

At this time, I was invited by General Pershing to accompany his party on a tour of Italy. They had visited England while I was at Fiume. We went first to Rome, where at a review the king[9] decorated us, awarding me the Commander of the Crown. We lunched with the premier. There, I met the minister of marine, who was very friendly. I told him that I wanted him to send Bartolucci to Washington, and he said that he would do so. We dined with the king, and I sat one removed from him. He talked to me across the minister between us much of the time. After dinner, when we repaired to a great reception room, the king came to me and said that Italy was much hurt by Mr. Wilson's policy of not letting Italy have food and coal. I had seen the suffering everywhere for lack of both. I told him that our foreign policy would probably change after the next election. I also asked him to send Bartolucci to Washington, and he said that he would.

While in Rome, we were taken to many historic spots and monuments. We then went to the battlefields, including the Asingo Plateau and the Piave River theater. We stopped overnight at Mantua and attended the open-air opera, which was free to all the people. At Milan, I was captivated by the cathedral, which was lace in marble. We went through the Brenner Pass and along Lake Garda through the Tyrol. I rode much with Mr. Jay,[10] the chargé d'affairs at our embassy in Rome. I told him that I thought the navy in the Adriatic felt slighted by receiving no attention or courtesies from the embassy. He said that he would correct the situation. Admiral Andrews later told me that the relations improved.

Chapter 22

Home and the First Division

Early in September, we received orders to return to the United States with General Pershing on the *Leviathan*, sailing from Brest. I obtained permission to bring my French aide, Lieutenant Gouin,[2] with me. Bartolucci told me that he had been appointed assistant naval attaché in Washington, but the Americans would not give him transportation on the *Leviathan.* I could not change the decision of the transportation. I told Bartolucci to go to Brest and to join me when our train reached there, which he did. I then asked the ship's quartermaster to let Bartolucci go on board. He said that there was no space. I told him to let Bartolucci share my suite of three rooms, and he consented. Not long after we went on board, Bartolucci said that Cromwell, who was on his staff,[3] wanted to share his apartment with him, and this was arranged.

Shortly after leaving, Major Lloyd Griscom,[4] who had been an ambassador and was then a major on General Pershing's staff, said that he was exceedingly worried about General Pershing's address to the joint session of Congress, which was to take place soon after we reached Washington. He said that General Pershing was out of touch with the way Congress and the people were thinking and asked me to prepare what I thought he should say. I felt that modesty on his part was all-important, and I stressed the credit due the troops and the support of Congress and the people in gaining the victory. Griscom liked it, but we continued to study and revise it so as to be brief and expressive. Before landing, he said that we should have the advice of some reliable man in Washington who was in close touch with public opinion. I told him that I knew Senator Elihu Root,[5] who was the ablest man in the country, and I would consult him.

We reached New York September 8, and I was met at the dock by my precious wife and son, whom I had so longed to see in the two years of separation. On leaving the ship, an order was handed to me assigning me to the command of the First Division at Camp Zachary

Taylor, Louisville, Kentucky. My cup of joy was full. Most officers on landing received orders demoting them to their permanent rank. We paraded up Broadway with the characteristic New York reception. There was a banquet that night. I sat next to General Wingate,[6] who commanded the New York artillery brigade of the Twenty-seventh Division. He said that on the night of the Armistice the brigade was resting in Verdun. They celebrated by sitting around campfires and singing Tobyhanna songs and telling Tobyhanna stories. No finer tribute could have been paid to the influence of that training camp in 1913–1916 on the war. The First Division had preceded us and paraded in New York on September 10. General Pershing and his generals rode at the head. I noted that the crowd, which in New York was always tumultuous in applause, was silent. Then I heard one man say: "Just look at their faces." The people were awed into silence by the grim faces of these fighting men who had returned as the miracle of war from the brink of death on every field.

On reaching Washington, I called on Mr. Root and showed him my notes for General Pershing's address. He said some things would not be wise and suggested other ideas, including the potential influence of the newly formed American Legion. When Major Griscom and I finished, I took the paper to Captain Quekemeyer, General Pershing's aide. That was the first intimation that they had of what we were doing. While our language was not used, our ideas influenced his remarks. I was among the generals who accompanied him to the joint session of Congress. Mrs. Summerall explained to Steamer that he could find a comfortable place to stay at the YMCA in Washington. The next morning, she asked him where he was staying, and he said: "At the Powhattan." Like his uniforms in Paris, he could not live below my status.

On September 17, the First Division paraded in Washington, and again General Pershing and his generals rode at the head of the troops. The ovation was overwhelming. The president and cabinet reviewed the parade in front of the White House.

On September 30, we left Washington for Louisville. On reaching the train, we found that the drawing room reserved for my dear wife and me was occupied by a young couple with a baby, who also had tickets to it. Always trusting of others, my dear wife said for us to give it up. By using Steamer's berth and procuring the other on the crowded train, we were accommodated. Our son had an appointment to West Point and was taking a preparatory course at

Schadman's School. He remained at the Westmoreland apartment where we had lived since 1914, under the care of General Mordecai, his grandfather.

On the train, I met Dr. Harrison Randolph,[7] who had been in the State Department during the war and was in charge of the affairs of the Balkan states and the Adriatic. He wanted to know about the Inter-Allied Commission of Inquiry on Fiume, of which the State Department knew nothing. He knew nothing of the London Pact between England and Italy on which Italy entered the war, or of the proceedings of the Allied admirals of the Adriatic during the war, or of the treaty between Italy and Austria, or of other important documents. I was able to furnish him with copies of all, including the report of the Inter-Allied Commission of Inquiry on Fiume.

We found a place to stay at Camp Taylor until we could rent a house. The division arrived promptly, and the troops were made comfortable in the camp barracks, which were superior to anything they had known during the war. The people of Louisville received us with open arms. Women's organizations provided dances and recreation for the men. We were entertained most lavishly at dinners and receptions. All were happy, and the morale was high. Many men wanted to be discharged, and there was delay in recruiting replacements. Training was conducted, and all were kept busy. The property was in chaos, and I devoted most of my time to having it inventoried and warehoused. Property and finance have always been of intense interest to me. What was my surprise to receive a few years later an inspection report from the War Department in which an inspector, by reason of some grievance that I never knew, recommended that I be held responsible for huge losses of which I could have known nothing. Of course, it was disapproved. I could later have injured this officer, but I never made reprisals on my enemies.

As soon as possible, I undertook to keep my pact with our dead on the battlefields and erect some worthy memorial to them. In this, I had the enthusiastic support of everyone. We organized the First Division AEF Memorial Association, and I was made president. We solicited funds from the division and collected $5,000. Our goal was $150,000. A circus was then proposed and soon became a reality. Under the management of Major Harcourt Hervey,[8] it included exhibition and spectacular drills by the infantry, artillery, and cavalry. He procured lions, elephants, clowns, etc. Seats for several thou-

sand people were constructed by Captain Vaughn,[9] who attended
to their erecting and shipment. We gave some trial performances
at Camp Taylor. The first night, a lion attacked and nearly killed
his keeper. This brought great publicity. The second night, the seats
were jammed. During a bloodcurdling attack of the tanks supported
by a terrific fire of blank cartridges from a battery of artillery, a caisson blew up, setting fire to the clothing of several men around it.
They ran with the fire blazing around them. Men threw them down
and smothered the flames. Terrible as it was, no one received any
permanent injury. The excitement of the crowd and the publicity
throughout the country made it famous. People came to see the lions kill the keeper and the cannoneers on fire.

Using three special trains, the circus went to Indianapolis and
drew capacity crowds for a week. It then went to the Lake Shore in
Chicago and drew overflow crowds for a week only a short distance
from Barnum and Bailey's circus. The expenses were very great,
but we cleared nearly $100,000. A meeting in a theater in Chicago
brought about $15,000 in subscriptions. It was presided over by
General Dawes[10] and was attended by such leaders as Mr. Robert
T. Lincoln,[11] Mr. Armour,[12] the Pullmans,[13] and others. Mrs. Ronald
Lyman in Boston, whom I had known as a child and whose brother, Major Cortland Parker, was in the First Division, invited me to
speak to some of her friends in her drawing room. Mr. Sargent,[14] the
great artist, Dr. Lowell,[15] the president of Harvard University, and
the leading people of Boston were present. They left subscriptions
of about $1,500. While in Boston, my friend General Sherburne[16]
took me to call on Governor Calvin Coolidge. He said almost nothing, but the general assured me that he was glad to see me. Later, he
sent me a copy of his book *Have Faith in Massachusetts.* When he became vice president and president, I telegraphed him, and he wrote
cordial replies. Gentlemen whom I met on my trips would give me
$1,000. Some thirteen thousand people sent donations, some only a
few dollars, saying it was all they had. In about a year, we had over
$150,000. I had only one unpleasant reaction from an officer of Ger-
man blood who resented me asking the army to help memorialize
the dead of a regular army division. Many soldiers and other commands sent small amounts.

After seeking advice, especially from the librarian of Congress,[17]
who was the secretary of the Commission on Fine Arts, we selected
Mr. Cass Gilbert[18] of New York as the architect and Mr. Daniel Ches-

ter French[19] of New York as the sculptor. They were the leaders in their fields. I met with them in New York and told them that I wanted them to give their hearts as well as their heads to their designs. Mr. Gilbert said that only three forms of monuments were possible: the column, the arch, and the temple. I told him that I wanted a column. He was then engaged in the construction of a state capitol and other great buildings, but he agreed to design a monument for us. I told Mr. French that I wanted the figure of the monument to symbolize triumph, sacrifice, and gratitude. He said that it could not be done. I asked him to try.

In due time, they evolved the beautiful design of the First Division Monument in Washington, DC. Later, when I was stationed in Hawaii, I saw the picture of the exquisite figure of Victory. I wrote Mr. French that it was too voluptuous, but he said that it would not be so high up on the column. The victorious figure with the raised flag showed triumph, the outstretched arm in benediction showed sacrifice, and the profound expression on the face showed gratitude.

I also tried to find a suitable spot. Mr. Gilbert said that a memorial should be remote but not too remote. I examined places on the axis from the Capitol to the Lincoln Memorial, but they were all inaccessible. I searched the perimeter for some future center like the Theodore Roosevelt Monument site but found nothing. One day, I entered the car on the Seventeenth Street side of the old War Department Building to discuss the matter with the secretary of the Commission on Fine Arts. As we drove to the back of the War Department, I saw the vacant square between the White House, the War Department, and the Corcoran Art Gallery. It was the perfect site for the monument. When I told this to the secretary of the Fine Arts Commission, he enthusiastically agreed. The Fine Arts Commission approved the design and the site.

We were now ready to ask for the authorization of Congress. A bill was prepared and introduced in both houses. It was referred to the Joint Committee on the Library, which decided such matters. We at once encountered the bitter opposition of Colonel Sherrill,[20] who was superintendent of public buildings and parks. He won to his side the chairman of the committee, a senator from Connecticut,[21] who afterward committed suicide. While we were engaged in our memorial, a meeting of the general officers of the war took place in the War Department. I appealed to them to have every division erect a monument to its dead in Washington. I suggested constructing an avenue of monuments extending west from the State, War, and Navy Buildings and calling it by such a name as the "Court of the Triumphant Dead." I could arouse little response. Later, a small arch was placed in the park by the Second Division. Other divisions did nothing. One day, I went to the Senate and sent a page with my card to a prominent senator on the committee and asked to see him. He did not come, and, while I was waiting, a doorkeeper saw me and spoke cordially. When I told him that I was waiting to see the senator who did not come, he said, "General, he can't treat you that way. I am one of his constituents, and I will make him see you." He went to the senator, who came at once, but, when I appealed to him to help get the approval of the Committee on Fine Arts, he was very cold.

I was made a major general in the regular army on April 20, 1920. The telegram notifying me was delivered to me at a dinner in Providence, Rhode Island, by my friends of the 103rd Field Artillery.

During our stay at Camp Taylor, many important events occurred. I had secured the appointment of our son, Charles Pelot Summerall Jr.,[22] to West Point by the Honorable Fred L. Blackman[23] of Alabama. I had made a friend of him while negotiating the purchase of the artillery range at Anniston. My son had remained in Washington to attend the preparatory school when we went to Camp Taylor, but he joined us during a vacation. We went to Washington to see him start for West Point early in June 1920. When we saw him go through the iron gate of Union Station to the train, I realized that he was going into the world alone and that I could never do for him anything that I had not done. We were proud of him and his opportunity, but our hearts were crushed at this definite separation. He amply rewarded us by his character, intellect, and achievements.

On one of my trips to Washington, not long after I had gone with Commander Bartolucci to see his fiancée at the Argentine embassy, I was told that he was ill in the Eye, Ear, Nose and Throat Hospital. I found him in agony with double mastoiditis.[24] His fiancée was with him, and they presented a tragic picture. I could do nothing. In a few days, I learned that he was dead. I have often wondered how much I was responsible. If I had not secured his detail to Washington, he might never have contracted the fatal illness. I had learned that his leaving me in Fiume was because he wanted to be with this young lady in Paris and his wanting to come to Washington was to be with her there.

Chapter 23

Camp Dix, New Jersey

The First Division was transferred in September 1920 to Camp Dix, New Jersey, which I reached on the twentieth. In Louisville, hundreds of the officers and men had married Kentucky girls. But there were no quarters for families at Camp Dix, so our first task was to convert a number of barracks into family apartments. This was done by the First Engineers. The New Jersey schools refused to receive the hundreds of children in the division. We gave entertainments, having vaudeville companies from New York, raised money, prepared schoolrooms, employed teachers, and conducted our own schools. The people of the area, especially in the towns, were hostile to soldiers, and it became necessary for me to put certain towns off limits because of venereal disease. This resulted in a suit against me for large damages, which failed. A commander has authority over and is responsible for his troops wherever they are.

As soon as we were settled at Camp Dix, I began writing a history of the First Division.[1] A board had collected some data, but it was not suitable for a history in either form, language, or substance. Sergeant Glidden[2] of the staff was an unusual artist and made some beautiful drawings for it. I sent a hundred mimeographed copies of my draft to officers and men who could check all of the facts. The result was in accordance with the knowledge of eyewitnesses. I soon saw that I could not mention the names of all officers and men who had performed outstanding services. The list would have been too long, and no doubt many of equal merit would have been omitted. The publishers required payment of $25,000 for five thousand copies. The funds received from the liquidation of the canteens in Germany exceeded this amount. The division judge advocate ruled that they could be spent for this or any other division activity. Before I left Camp Dix, we ordered an edition of five thousand copies. There was no demand for them by the public, and the response by our men was disappointing. Many copies were given away. I did not take credit as the author because I did not want to appear to be seeking distinction, but I compiled it and wrote it entirely. I consider it an epic as a story, and the English, imagery, and sentiment are superior. Perhaps it went over people's heads.

On Cantigny Day, May 28, 1921,[3] the division gave a pageant. Arches representing the different battles and sectors were erected, and the troops passed under them. Many veterans came, and we had a meeting of the Society of the First Division. General Pershing spent the day and the nights preceding and following. He showed every evidence of affection and gratitude for the division. During this time, he made himself one of us.

General Douglas MacArthur, who was the superintendent at West Point, asked me to let him bring the first and third classes of the Corps of Cadets to Camp Dix for summer training with the First Division. We prepared a camp and arranged for them to have target practice and maneuvers. We all considered the experience very profitable for them. My son was a member of the third class.

We found Camp Dix in a badly deteriorated condition. So we repaired buildings, rebuilt the railroad to the target range, and improved living conditions. My dear wife made our house the stopping place of the families who arrived and gave shelter to young mothers, children, and the sick. She and other ladies organized a welfare group to visit the soldiers' families and see that their needs were supplied. We were visited by the daughter of General di Robilant, who was with me at Fiume. A number of Philadelphia ladies brought entertainment to the camp and invited the officers and men to entertainments in Philadelphia. General and Mrs. MacArthur invited my dear wife and me to spend a weekend with them at West Point, where he met us with our son in his car. He also asked me to talk informally to the cadets.

I had arranged with the War Department to have First Division officers and men detailed to the duty of compiling the records of the First Division in the War Department. This was done thoroughly, and one hundred sets of the volumes and maps were prepared and distributed.[4] We found a large number of prisoners at Camp Dix who gave much trouble. There was a residue of lawless men who bootlegged liquor and stole property. During the year, recruits were received. The division was well trained, and, in spite of the isolation and poor living conditions, morale was high.

Chapter 24

Hawaii

I was ordered to the command of the Hawaiian Department, and my dear wife and I, accompanied by my two aides, Lieutenants Forster[1] and Giles,[2] left Camp Dix June 30, 1921. I was succeeded in command of the division by Major General David Shanks,[3] who had commanded the port of embarkation at Hoboken, New Jersey, during the war. As soon as I left, he ordered all the records, photographs, and valuable documents in the offices of the Society of the First Division and the First Division Memorial Association destroyed. He then, I was told, stopped all these activities and spent his time playing golf.

I assumed command of the Hawaiian Department August 5, 1921. I was fortunate in finding as chief of staff Colonel William Chamberlaine,[4] a classmate whom I had always admired and who was one of the most loyal and capable officers in the army. Our baggage went with us, and we were soon settled in the spacious commanding general's house at Fort Shafter. We at once secured a capable Japanese cook and his wife as servants. Colonel Chamberlaine assured me that we would never lack for servants, as the best spies would be supplied me. I have always thought that this cook was a Japanese officer of important rank as he was too artistic and intelligent to be a servant. We were careful never to have any papers or to discuss official matters at the house. Hawaii captivated us by its charm, and the social life began at once and continued throughout our stay.

I began visiting the troops of the Hawaiian Division at Schofield Barracks, the coast artillery in the fortifications, the air corps at Ford Island in Pearl Harbor, and the quartermaster's depot. When I talked to the men, I sensed a lack of response and goodwill. The dress of officers and men was poor and the saluting and military bearing low. I held conferences with commanding officers and staffs every week. By orders and personal appeals, I tried to bring about a more soldierly dress and bearing and was soon successful. I ordered a training program to bring the troops up to a high standard of efficiency. I impressed on them that we were in the front line and must be ready for an attack by Japan at any time.[5]

Before leaving the States, I was told in Washington that the Japanese had considered an attack before the 1920 disarmament conference[6] in Washington and that this, together with the unwillingness of England, which was in alliance with Japan, to finance such a war, caused Japan to delay. The G-2 [Intelligence] Section had much information on Japanese espionage. Japanese ships transporting oil, scrap, and supplies from the Pacific coast passed weekly with different crews and alternating calls at Honolulu and Hilo.

There were no war plans, and the situation was tense. I directed Major Lesley J. McNair,[7] the Operations Officer (G-3), and Major C. H. White,[8] the assistant G-3, to seclude themselves and devote their entire time to the preparation of war plans. Commander Chester Nimitz[9] of the navy was in command of the dozen old submarines in Pearl Harbor. I unofficially attached him to my staff, and he collaborated with Majors McNair and White. The study soon showed that the 15,000 troops in the command were woefully inadequate. It was found that at least 100,000 troops would be required to protect the potential landing places around Oahu. No effort was made to hold the other islands, as Pearl Harbor was the only strategic point to protect. The plan was for the army to defend Oahu. The navy was to seek the enemy's fleet and defeat it without any responsibility for land defense. We learned privately from Washington that the navy planned to assemble the fleet in Hanalei Roads for operations. To us, this would expose the ships to attack by enemy submarines and was unthinkable. At that time, planes could not fly safely from Oahu to Kauai, a distance of one hundred miles.[10]

Our war plan provided for defending the perimeter of Oahu with machine guns and beach-defense 75-millimeter guns. The railway guns were located so as to reach enemy transports at anchor. The fixed defenses and twelve 240-millimeter howitzers in permanent positions were also to attack transports and accompanying warships. Positions were selected for machine guns so that at least two could bear on every point of approach to the shore. Locations were also selected for ten 75-millimeter guns so that at least two could fire on every approach. All men were to fight in place and not withdraw in the face of a landing. We had an abundance of mustard gas to use against the enemy's ships. We considered that he might lay down a smoke screen to cover his approach. A secured line of defense was prepared behind the beach-defense line to stop him if he penetrated the first line. The north shore of Oahu was the most vulnerable, with the west shore also a practicable landing spot. The east shore was covered by extensive coral reefs, but holes in the reefs permitted the passage of landing boats. The south shore was defended by the fortifications, and transports would have been subjected to heavy artillery fire. Reconnaissance was to be continuous by small boats with radio and by the planes within one hundred miles of the shore. We expected to discover the approach of an enemy force at least the day before it could arrive at an anchorage. We knew that the Japanese aliens in Honolulu would keep the enemy informed of our condition and our movements and would meet him some hundreds of miles away in fishing boats. We expected the enemy to anchor during the night, begin bombardment of the shoreline at dawn, and start landing parties from the transports. We expected at least two separate serious landing attempts on the north and the west coasts and feints at landings on the west and south coasts, which would become effective if we could not repel them.

It was inconceivable that any American ship would be in Pearl Harbor except the two station destroyers and the twelve old model submarines, which would at once attack enemy ships. Any fleet in Pearl Harbor could be blocked by one of the Japanese ships that passed near the channel every day or two. Within a half hour, a Japanese ship could enter the channel, sink itself, and prevent any navy ship in the harbor from leaving for weeks until the obstruction was removed. Airplanes could not fly far over water, and we would have discovered a carrier force before it approached near enough to have planes attack. We also knew that ships in Pearl Harbor would be attacked by enemy submarines as they tried to come singly out of the narrow channel. We not only considered having no navy ships in Pearl Harbor; we also did not want any part of the navy in Hawaiian waters, where it would be exposed to submarine attack without warning. Moreover, land defense was against all navy policy and prevented the mission of the navy to seek the enemy at sea and defeat him, thus accomplishing the best form of defense. In accordance with the principle of paramount interest, the command of the defense of Oahu devolved on the army. Such naval forces as were there would pass under the control of the army. Pearl Harbor was to be only a harbor of refuge for damaged warships after naval combat.

We expected the Japanese population to cooperate with the enemy at the time of attack. As soon as invasion became imminent, all dangerous leaders were to be arrested. Cordons of troops were to separate the Japanese population and prevent assemblies. We had an X alert and a Y alert. The X alert required all troops to move to their battle positions. Trails were cut from Schofield Barracks to the east coast and to the Pupukea Plateau, defending a large part of the west coast. Also, a road was made practicable from Schofield Barracks to the west coast. We expected to declare martial law, and the proclamation and orders were prepared. The territorial authorities were to control the civilian population.

In order to provide water for a sufficient garrison, a large reservoir was constructed above Schofield Barracks, and a refrigeration plant was built at Schofield Barracks to hold meats and perishables for 100,000 men for six months. Maneuvers and field exercises were constantly held day and night. The troops were familiar with their missions and could be relied on to reach their posts quickly. I have never seen better-trained troops or higher morale than were developed. At my inspections, colonels would ask me to call on them for any kind of tactical requirement, and they were sure that their regiments could do well.

Every means was adopted to raise morale. I wrote letters of commendation and spoke in a complimentary way to individuals and units whenever warranted. In my annual inspections, I had commanding officers present their subordinate officers and men who had made the commands better by their presence, and they were cited before the command. At one inspection of the Hawaiian Division, I noted a reel cart [of signal wire] in the most perfect condition. The horses were beautifully groomed, the harness highly polished, the cart painted and varnished, the reel and wire looking like new, and the driver in a handsome uniform, sitting in a most soldierly manner. I stopped and complimented and praised him in the presence of the division and department staffs and in hearing of the troops. Within an hour, the chief of staff of the division[11] reported to me that the driver was dead. While unhitching the horses, one of them kicked him over the heart and killed him instantly. Although I tried never to find fault, at one inspection I found on entering a barrack that it had not been prepared, and I left telling the colonel that I would inspect it at another time when it was in order.

At first the men said that they could not clean the harness and vehicles because the volcanic dust could not be removed. I had every driver's name stenciled over his harness on the heelpost and on the end of his wagon seat. In a short time, all was in excellent condition. A man will not allow his name connected with anything discreditable.

To receive visiting admirals, a guard of honor from the 155-millimeter gun regiment of Fort Shafter was always detailed. The men all had tailor-made uniforms and presented a superior appearance. Vice Admiral Field[12] of the British navy asked why they wore such uniforms. I told him that they purchased them voluntarily from pride. He asked if it paid. I told him that the high morale was reflected in all they did.

The coast-defense artillery conducted all target practice at extreme range, and at times the target was over the horizon and fire was adjusted by a spotter in an airplane. The field artillery also adjusted fire by plane. Half of the coast artillery and heavy mobile artillery practice was conducted at night and half by day.

At first, Pearl Harbor was commanded by Admiral Simpson,[13] with whom I worked in the closest harmony. He was succeeded by Admiral McDonald,[14] who would have nothing to do with me. This was overcome by Captain Babcock,[15] the chief of staff, and Commander Nimitz, the commander of the submarines, who worked in the closest cooperation. They always accompanied me as a part of my staff when I inspected the troops.

After the supply, communication, and medical annexes were prepared for any emergency, our war plans were complete. We were visited by various inspectors from the War Department, who approved our plans and training. Brigadier General William Mitchell[16] arrived and stated that he came for a hunting trip. He made no inspection, and his presence was entirely unofficial. I had known him since he came into the service from civil life, and we gave him and his wife a dinner. He told me that he would like to fly one of the De Haviland planes [a DH-4 bomber] to Kauai. I made all arrangements for him to do so. He then said that he decided not to fly, asked me to send the plane to Kauai on a tug, and said that he would fly it there and photograph a small island near Kauai. This I did. When he tried to take off on Kauai, his plane hit a fence, and he did no more. I believed that he had no intention of flying and hit the fence on purpose. He left, and I thought no more of him. I had been with him at Fort Myer, in China, and in France. While I felt friendly to him, he had always seemed irresponsible, eccentric, and vain. I had no confidence in him or respect for his attainments. I regarded him as a showman. It will be seen later how he abused his presence in my command.

On arrival in Honolulu, I found the press very unfriendly, and there were daily attacks on the troops. Even socially, men and women told me of outrages that the soldiers were committing, I knew their statements to be untrue. I told the G-2 [intelligence officer] to arrange with the papers to publish news items as far as they would give space. In a short time, the papers became friendly, and the slanders of the troops ceased.

It was necessary to put a section of Honolulu called Tin Can Alley off-limits and to have the military police keep the men out. The men resented this at first and called Oahu *Summerall's Rock.* But it was necessary to find proper recreation for them. One of the editors, Mr. Thurston,[17] who was very unfriendly to my predecessor but became a good friend of mine, gave me the use of a tract of land near Kilauea Volcano on the island of Hawaii.[18] I had barracks and a mess built for the men, and the regiments constructed cottages for the officers' and enlisted mens' families. I placed a cadre in charge with a medical officer. The troops went there by companies and battalions for a week at a time. The climate was cold at an altitude of four thousand feet. The scenery was grand. No duties were required. They had complete relaxation and came back refreshed. We could get no help from the War Department, and we paid the cost of shipping supplies and the tickets for the people on the interisland boats. I had trucks and cars take them from Hilo the thirty miles to Kilauea. All families took turns in occupying the cottages and ate at the mess. The health, morale, and spirits became excellent.

The normal tour of duty in Hawaii was three years, but I let all officers have their accumulated leaves and go home before the full time. Men also were given furloughs to shorten their tour. Three years was too long for a tour of foreign service. Mental and physical deterioration had an effect on men, women, and children. A bathing beach with bathhouses was established at Haleiwa on the north coast, six miles from Schofield Barracks, and the men went there in trucks. The children at Schofield Barracks could not go to the Hawaiian schools. I provided buildings, and the wives of officers who had taught were employed as teachers and paid by the territory.

All forms of athletics were developed to a remarkable degree.

Boxing was the most popular sport, and the matches between the army and the navy drew not only the services but thousands of civilians. I had our boxers do nothing but train so that, when they entered the ring, they could take care of themselves. Some became professional after discharge. The football competition was nearly as keen. The navy would collect their best players from the entire service after the season ended in the States and send them to Pearl Harbor to play the army during the Christmas season. Of course, they won. In my last year, I formed the football players into a company, and they did nothing but train. Only one officer was allowed on each team. Lieutenant Sproul[19] had the army team. The game was close, but in the last minute army made a touchdown that won. I went to Admiral McDonald to congratulate his team, but he would not speak to me.

One dramatic event during our service in Hawaii was the death of the last prince of the Hawaiian royal family.[20] It occurred while Admiral Simpson was in command at Pearl Harbor. The body lay in state for a week in the old Hawaiian church with watchers from the *kama'ainas*, or old inhabitants, gently swaying the picturesque fans. All flags were kept at half staff during the funeral procession. I had a battery fire minute guns from the top of Punch Bowl, and the royal salute was fired at the end of the commitment services, with volleys and taps.

When the territory was incorporated into the United States, the crown lands passed to our government. They afforded gun sites, campgrounds, and maneuver areas, although all that were suitable were leased for sugar cane plantations. I approved the exchange of some sand pockets on the west coast for some campsites on the north coast. I used the old arsenal for a recreation center in Honolulu. The troops pitched tents and aided the Boy Scouts to have camps. This was an important service to the organization.

Warships from Great Britain, France, and Japan visited the Islands. The British Special Service Squadron consisted of seven British and Australian cruisers, with the HMS *Hood* as flagship, commanded by Vice Admiral Field. Before arriving, he radioed to ask me if I preferred not to have liquor served. Under Prohibition, I replied that I would rather he did not serve liquor. Usually, the guests drank excessively on visiting ships. He did not serve any liquor. A series of dinners took place of which we gave a large one.

A Japanese naval squadron also came with two young royal

princes on board. Of course, we entertained them. At the dinner given by the Japanese consul general, Mrs. Summerall found that the princes wanted to dance but were not allowed to do so. She asked the admiral to let them dance with the young daughters of the consul general, which he did, though, I thought, reluctantly. I always furnished these naval visitors with cars, aides, and horses to ride.

Our social life in Hawaii was very active. We either entertained at dinner or attended dinners much of the time. We were at home once a month, when hundreds of army people and a few civilians came. When the Japanese earthquake occurred,[21] the army sent from Honolulu a shipload of supplies. Many prominent people visited the Islands and were entertained by my dear wife and me. Her grace, charm, and kindness captivated everyone. Guests lingered at our receptions and dinners as though loathe to leave. I believe that neither there nor elsewhere where we were stationed was the army ever so well represented.

Our dear son came to us for his furlough in 1922 and took an active part in the water sports and social life. We enjoyed riding together. His mother went to see him graduate from West Point in 1924, and they returned to Honolulu for his leave.

When I went to Hawaii, the efforts to secure approval of the Joint Committee on the Library for the First Division Monument were continued by the Monument Committee. Colonel Clark Williams[22] was especially active. One day, they went to a hearing by the committee, of which Senator John Sharp Williams[23] was a member. When he entered, he asked what the meeting was about. On being told that the First Division wanted approval of a bill to erect a monument, he asked if they had the money. When told that they had the money, he exclaimed, "Then approve the bill," which was done. The bill passed both houses, but a German congressman from Ohio[24] amended it in the House by providing that no federal money should ever be spent for erecting or maintaining the monument. The work of erecting it then proceeded. The publication of the *History of the First Division* also proceeded, and I received a copy of the first edition in Hawaii in 1922. The volume was very handsome and the contents worthy of the great events recorded.

Chapter 25

Return to the States

After three years in the islands, the War Department issued orders relieving me of command of the Hawaiian Department and ordering me to assume command of the Eighth Corps Area with headquarters at San Antonio, Texas.[1] About a month before the expiration of my tour of three years in Hawaii, I began receiving letters from a man in San Francisco who said that he was a veteran of the Sixteenth Infantry of the First Division in the World War. He wanted to arrange a welcome for me on arrival in San Francisco. I could not refuse but had no idea of what was to take place. On leaving Honolulu, many courtesies were extended to us, including a meeting of the Boy Scouts in a theater where a handsome cane was presented to me. Other trophies were presented by individuals or organizations to my dear wife or me. The papers published appreciative editorials on my service and accomplishments. We sailed from Honolulu on a transport on August 12, 1924, and were covered with leis from many friends who came to the boat.

On reaching San Francisco, we were met by a delegation of the city council and prominent men and by a committee of ladies with flowers for my dear wife. A police escort conducted us to the Fairmont Hotel, where a beautiful suite was reserved for us. The mayor and the city council received me officially and assured us of a warm welcome. The lunch clubs and the city council gave a united lunch, where I spoke. My theme was the unstable condition of the European governments and the heavy arming of the dictators while we reduced to a skeleton army of 118,000 men. I explained offensive and defensive armaments and showed that offensive armaments in Europe and Japan were for aggressive war, in which we would be involved. I warned of the heavy armament of Japan and the danger to the Pacific. My talk was a prediction of what eventually took place and a warning as to what we should do. Of course, what I said was not understood or taken seriously, there or anywhere else.

As I had had no leave since the war, I took three months leave, and we spent most of it with friends in Santa Barbara, California. Arrangements had been made to unveil the First Division Monument in Washington on October 4, 1924, the anniversary of the division's entry into the Meuse-Argonne battle. I felt that my address was to be significant, and I spent much time preparing it.

My dear wife and our son preceded me on the trip east, as he had to report to his regiment, the Sixth Field Artillery of the First Division at Fort Hoyle, Maryland, by September 30. I joined my wife in Washington. The dedication ceremony and unveiling of the monument were memorable. The First Division paraded along Pennsylvania Avenue. A large crowd assembled at the monument, where chairs were provided. The troops formed around the monument. President Coolidge and I were the only speakers, while a choir from one of the churches and bands of the First Division furnished appropriate music. The arrival of the president was timed to follow my address. What we said was afterward published in a beautiful volume by Mr. James Simpson of Chicago.[2]

After the dedication of the monument, my wife and I went to San Antonio. The entire Second Division met us when we arrived at dawn on October 6 and escorted us to the headquarters. Although I assumed command, it was understood that I would succeed General Bullard in command of the Second Corps Area with headquarters at Governors Island, New York, when he retired in January 1923.

We entered on the life at San Antonio, however, as though it were permanent. I made an inspection of all the posts to become acquainted with the troops and conditions. The command was very active and the troops in a high state of efficiency. The Second Division at Fort Sam Houston was commanded by Major General Preston Brown,[3] who was one of the most efficient officers in the army. I had known him since I helped coach him for a commission. We had long been close friends. We gave a reception for the officers and their wives. Several hundred came, and my dear wife was able to identify them with their names after being received and to introduce them to one another. The social life was very attractive.

On leaving one post after a visit, I told the commanding officer to requisition stoves at once for the quarters because some of the families were cold. I had visited a sick officer and learned that the stove in his room had been sent to the commanding officer's quarters for my room. The commanding officer afterward told me that he did not know that there were no heating stoves on the post and marveled at my thoroughness.

A notable event during this time was the death and funeral of Mr. Samuel Gompers,[4] the president of the American Federation of Labor (AFL). He had been responsible for the great railway strike in 1894, and I had been deeply prejudiced against him. During the world war, he had controlled labor, no doubt because he was an Englishman, and we were helping England, and there were no important strikes.[5] I had given him credit for this and considered that he had somewhat redeemed himself. One day, he arrived in San Antonio with the vice president of the international unions and a large party from a visit to Mexico. It was reported that he was very ill at the Saint Anthony hotel. I sent an aide to offer my sympathies and services.

He died during the night, and the next morning I sent the chief of staff[6] to offer my condolences and assistance. To our surprise, they asked for a military funeral. The chief of staff pointed out that it was not authorized by army regulations, but I told him that I would order the funeral. He urged me to refer the request to Washington. I declined and told him that I would take full responsibility. I saw an opportunity for the army to remove some of the hostility toward it by organized labor. I called General Brown and told him to make arrangements for a great funeral. He saw the advisability and went about it with enthusiasm. The body was borne to the train at midnight on a specially built catafalque, with a large military escort and with the entire Second Division lining both sides of the streets from the hotel to the railroad station. The bands played the funeral marches, and the display was almost barbaric. Of course, the labor officials were delighted. The publicity was enormous throughout the country. I traveled on the train during the night on an inspection trip, and my aide sat with the labor party. As a result of the publicity, a guard of honor met the train at every place where troops were stationed en route to Washington. At Washington, a guard of honor met the train, the casket was borne to the Capitol on a caisson, and soldiers stood watch while it lay in state. At New York, a guard of honor met the train, the casket was borne to the grave on a caisson, military salutes were fired, and taps was sounded at the grave. The value to the good relations between the army and organized labor was incalculable. During the rest of my service, labor officials were friendly to me and to the army.

On going to Fort Sam Houston, I found an officer on the staff whom I had been compelled to relieve from command in France.[7] He soon came to me and told me that when I relieved him, he debated for a long time whether to shoot me or to shoot himself. He decided not to do either. He wanted to tell me that I was right in relieving him and that he had no feeling against me. He assured me of his loyalty, and I enjoyed his friendship afterward.

Another officer with the troops had been unfriendly toward me because I had tried to discipline and train a battalion that was in bad condition at Fort Myer, Virginia. He also came to me and expressed regret for his disloyalty and assured me that he realized I was right in my methods and that I could depend on him for fidelity and cooperation. I have had this happen in several other cases. This officer became a major general in World War II. His moral courage and good sense in seeing his error showed his ability.[8]

Chapter 26

From Texas to New York

Pursuant to orders of the War Department, on the retirement of General Robert L. Bullard, I left Fort Sam Houston on January 11, 1925, and assumed command of the Second Corps Area with head-quarters at Governors Island, New York City, on January 16. At a large banquet for General Bullard in New York, I tried to pay suit-able tribute to him.

My dear wife used money that she had inherited and bought rugs and furniture for the large house of the commanding general. This was the most important command in the army. I visited the posts in New York, New Jersey, and Delaware and also called on the state governors within the corps area, as I had done on arrival in all of my other commands. We found constant social demands, and, generally, we went out to dinner or entertained at home. At the banquets of large organizations, I met and made friends with many prominent men.

A banquet was given to me as a welcome. Dr. John Findley,[1] prominent in education and journalism, was called on to make the welcoming address. He said that he knew nothing about me but would tell me about New York. He gave a beautiful picture of the human side of the great city. We became good friends. One of the most representative banquets was given to President-Elect Mucha-do[2] of Cuba. I was the principal speaker, although I did not know that I was to speak until I saw the menu. We had dined the previ-ous night with a prominent couple where the wife's first husband had been consul in Havana. She told me of the Cuban psychology and the price of Spanish blood and tradition. I used this informa-tion and so pleased the president-elect that he rose and shook my hand, thanked me effusively, and asked me to send him a copy of my remarks for publication. At a large banquet of the Iron and Steel Institute, I sat next to the president, Judge Elbert H. Gary.[3] Again, I did not know that I was to speak until I read the menu and saw that the British ambassador[4] and I were the only speakers. I told them of the relationship of national defense and industry. It pleased them, and several thanked me, saying they did not like to have the ambassador read to them. I always spoke extemporaneously.

I was made an honorary member of all the New York clubs. I was especially impressed by the Chamber of Commerce of the State of New York, where I spoke and pointed out the necessity for capital as an element of national defense and the need for commerce to be incorporated into war plans. One gentleman expressed some alarm when I said that everything in the country was a part of war and not just soldiers and sailors. He asked me if people's money was included. I told him that it was and must be used if the government needed it. It was no more important than the lives of men.

The troops at Governors Island were badly housed, but I could get no money for new buildings. But when the barracks burned one night, I went to the members of Congress, and over $1 million was appropriated for new barracks. Although I had no control over the Brooklyn base,[5] I learned of bad conditions as to property there and the treatment of men to be discharged, which I could not correct. The posts, mostly garrisoned by the First Division, were in a high state of efficiency and morale.

I conceived the idea of moving the troops from New York Harbor by barges, as we had no money for rail transportation, to Fort Ontario, overlooking Lake Ontario at Oswego, New York, and holding maneuvers at Pine Camp Reservation.[6] I knew that a large sum of money had been spent by the state to improve the Erie Canal. I learned that the highway bridges over the canal had only twelve feet clearance and only the specially built barges and tugs could pass under them. At a dinner shortly afterward, I sat by the wife of a man in the shipping industry. When I told her this, she said: "Yes, and my husband bought all of the canal barges and tugs." There was no other traffic on the canal. At another dinner given for me by Mr. John D. Rockefeller, I sat between him and Mr. Crowley,[7] the president of the New York Central Railroad. It was proposed to deepen the Hudson River so that large ships could go to Albany and to establish the Port of Albany. I asked Mr. Crowley what the New York Central thought of the plan. He said that the New York Central was giving $10 million in port facilities to get the project because anything that brought business into the territory of New York Central brought business to the railroad. This was a repetition of the experience in Hawaii, where increased facilities brought increased demand.

At various times, I was called to Washington on classification and selection boards.[8] At the court-martial of Brigadier General William Mitchell, by seniority I was president of the court. When the court met, he at once objected to me on the ground that he had made an inspection of Hawaii when I was in command and had reported to the War Department that I had no war plans. He had made no inspection, and when he was there, our war plans were complete. I had no recollection of his making this report. But, of course, I was excused from the court-martial.[9] The War Department cabled for the war plans, which came and were placed in evidence to show his falsehood.

Among my close friends was Governor Al Smith[10] of New York. At a dinner given me by the chamber of commerce in Albany, he spoke very kindly of me. On leaving early, he asked me to go to the executive mansion when the dinner was over. I found him alone, except for the chief of the state constabulary waiting for midnight, when some men were to be executed in Sing Sing. He said that he was deluged with appeals for delay, which he could not grant, and he wanted me to be with him until the executions were over. We talked until a report came of the executions, when I left. When he was the candidate for president, he told me that if he was elected, he would make me secretary of war. I regarded him as one of the ablest men I ever knew.

When the bust of General Sherman was unveiled in the hall of fame at New York University, I was invited to make the address. General Pershing and other distinguished men were present. I drove him and an ex-ambassador to and from the ceremony. Mrs. McKinlock, the mother who was our guest at Cochem Castle in Germany, gave a student dormitory to Harvard College, where her son had graduated. She asked me to make the address of dedication, but the president, Dr. Abbott L. Lowell,[11] ignored my presence and showed me no courtesy. Major General Preston Brown, who was in command in Boston, joined me. I saw Dr. Lowell several times in New York, and he always seemed to be inadequate.

When the AFL invited me to speak at their halls and on Labor Day, I offered them the use of Fort Hamilton for their meeting. Several thousand were present when I spoke to them. There was much criticism in labor organizations of the Citizens' Military Training Camp (CMTC)[12] program. Many were opposed to any military policy and regarded the army as an enemy. I invited President William

Green[13] and his international vice presidents to visit Plattsburgh, New York, and see the CMTC. I told them that I would have cars meet them in Albany and drive them through the beautiful Lake Champlain country to Plattsburgh. They could be guests of the officers or stay in camp, as I would do. I received them at Plattsburgh, where they reviewed some thousands of boys in the CMTC. That night, the local union gave a large banquet. I gave a welcoming speech and explained briefly about the camp. The next speaker was a leading international vice president and head of one of the largest unions. He at once assumed a hostile attitude. He said: "I have no roses to throw at the CMTC. I see no good in it." At once, a tension seized the audience. The room became deathly still. No doubt, everyone was shocked. The man stopped and remained silent for what seemed a long time but was no doubt only a few seconds. His manner and appearance changed, and he said: "I wish there had been a CMTC when I was a boy so that I could have gone to it. I wish I had a boy to go to it." Then he paid the organization the greatest tribute. He sensed the resentment to what he came prepared to say and was converted on his feet. Mr. Green spoke eloquently of the movement and paid the highest tribute to the army. From that moment, the AFL was a friend to the army. We had no more opposition. In all the years that followed, Mr. Green never failed to send me a handsome Christmas card. We included him as a guest at one or more of our dinners afterward in Washington.

One Sunday afternoon, I received a report that Picatinny Arsenal, New Jersey, had blown up. Within a few minutes, cars were speeding with men from Governors Island to the disaster. Other men with food and camp equipment followed. It appeared that most of the arsenal was destroyed. Because the employees were absent, there was no loss of life. The secretary of war came at once, and I drove him to the scene of the disaster. The troops took charge and remained until the employees could clear the debris and care for the property.

Each summer, the governor of New Jersey invited my dear wife and me to visit him and his wife at the governor's cottage at Sea Girt when the state national guard was in camp. These visits were most agreeable, and I was able to contact the state troops, many of whom I knew, in a most successful manner.

The management of the Metropolitan Opera House placed a box and some seats for the opera at our disposal each season. The different families and individual officers at the harbor posts were able to occupy them. We also enjoyed the friendship of Major Bowes,[14] who conducted the most popular radio program, and his wife, the former distinguished actress Margaret Illington,[15] brought us beautiful presents when she returned from abroad. Another benefit of corps area command was that my car had the right of way in New York and could pass all lights with police cooperation.

I was invited to the Grand Army of the Republic (GAR)[16] parades and to lunch with a small group of Civil War veterans who passed away entirely during my stay. I also attended a GAR Convention at Saratoga and spoke. We were invited to many large events, like the Danbury, Connecticut, Fair, where we made numerous friends for the army. Of course, courtesy calls were frequent with flag officers of our navy and the commanding officers of foreign naval vessels.

The training and morale of the troops were superior, and I felt at home with the First Division occupying most of the posts. In my inspections, the men who served in Europe were assembled at each post, and I spoke to them. This in no way antagonized the officers and men who had joined later. The people in the towns near the posts were very friendly and often entertained me. The old posts along the Canadian border were of historic interest and showed that Canada had not always been friendly.

An event of the greatest importance in our lives was the marriage of our son to Julia Reeder[17] at Fort Monroe, Virginia, October 23, 1926. My dear wife and I went for the wedding and were most cordially entertained by General and Mrs. R. Callan.[18] He commanded the post and was an old friend. The wedding in the post chapel was very beautiful, and the reception in the club rooms in the old fort was perfect. It was a happy beginning of a happy life for them and for us in them. We did not lose our precious son, but we gained a darling daughter.

Chapter 27

Chief of Staff

The announcement of my appointment by President Coolidge as chief of staff of the army was received by the press, the country, and the army with general approval. While I had never felt ambition for advancement, my dear wife and I were deeply gratified. Many letters came from old friends, and editorials in the papers were very complimentary. I relinquished command of the Second Corps Area on November 20, 1926, and assumed the duties of chief of staff November 21. We stayed for a few days with our good friends Mr. and Mrs. A. B. Butler[1] in their beautiful new home. However, we were soon settled in the chief of staff's home at Fort Myer, Virginia. The first and most urgent duty was to familiarize myself with the budget, which had been prepared for the next fiscal year. Many matters in the War Department arose for study, decision, or recommendation. Social demands at once claimed most of our evenings.[2]

The War Department General Staff and chiefs of branches had nearly all been a part of GHQ in France. Some of them, including the deputy chief of staff,[3] had already secured orders assigning them to other commands. I realized that they held the prejudice of GHQ against me as being only a "combat officer," and I did not expect full loyalty from them. No doubt, they resented my selection when one of them especially felt a claim. The General Staff law specifically exempted the chiefs of supply services from the control of the General Staff and placed them under the assistant secretary of war. From the beginning, this official[4] and his large office staff were unfriendly to me. For a time, I continued the weekly staff meetings of these chiefs and the General Staff heads in my office. It was manifest that they came reluctantly and, no doubt, felt that they need not come at all. I, therefore, discontinued these meetings. The replacements for the General Staff sections with those who remained and the new deputy chief of staff[5] were friends and were loyal. The situation required tact and firmness within my authority and patience. I tried to avoid an open break with the chiefs of supply services, but this was only deferred with some of these individuals.

The greatest evil in the War Department was the continued re-appointment of chiefs of bureaus, contrary to the intent of the law, which said that they should be appointed for four years. As terms of some were soon to expire, it was necessary for me to deal with this question promptly. It caused resentment in the army and deprived good officers of any chance for promotion as chief of branch. I soon discussed it with the secretary of war,[6] who agreed that such of-ficers should not be reappointed. He consulted the president, who approved of our decision. My policy in some way became known, and one chief of branch served notice on me, naming others who would help him, intending to fight me and discredit me in every way possible. They had built up a good deal of political influence with many friends in Congress. Although I had been instrumental in securing the appointments of some of these men and had regard-ed then as my friends, with one exception all became my enemies. The practice was age-old, and no one had ever dared to oppose it. Theodore Roosevelt stated in his book[7] in 1899 that it was the cause of inefficiency in War Department administration. None of the hold-ers of offices were reappointed, and the policy remained unbroken.

Another evil that required prompt attention was the redetail of officers for duty in the offices of the chiefs of branches and on the General Staff after they had been away from Washington the two years required by law. Many officers in the army were able and were entitled to these details. I stopped the practice at once and thus made many more enemies. The result, however, was a correspond-ing rise in morale of the army and a feeling of justice and opportu-nity. Unfortunately, only a few officers could be selected, and many who wanted the details became resentful and hostile. The same feeling resulted from the selection or promotion of general officers. As a friend said: "With every appointment, I made one ingrate and twenty enemies." I brought some brigadier generals to the General Staff sections who had never been allowed to come. I knew that they would become major generals and that they should have experience in the War Department. I assigned to one the duty of visiting the entire army and interpreting the War Department to the army with a view to removing the strong prejudice against the War Department.

In the meantime, hearings on the budget by the subcommittee on appropriations took place.[8] I was familiar with the large volume of items and explained them personally. This pleased the members of the committee, and my relations with them were always cordial. At the same time, the House Committee on Military Affairs had hearings on requests for authorizations. We were especially in need of housing. On the first day, I listened to the committee treat the secretary of war rather harshly. When I appeared the next day for my presentation, I asked the committee to allow me to tell them what I considered the army to be in order that they should know what I was talking about. I described the kind of people who constitute the officers and their families and the type of men who made up the enlisted personnel. I explained their lives, their limitations of action, and their dependence on the government for fair treatment, decent homes, and the whole procession of their lives. The committee listened attentively. When I had finished, one member said: "General, I wish every person in the United States could have heard you." I then presented the requests for authorizations and was treated with every consideration and courtesy. My relations with the committee were always cordial and friendly. Two members came to my office at different times, saying that it had been years since they had gone to the chief of staff's office and that they never intended to go but that I had changed their feelings. I knew some of the members of the Senate committees, and my relations with the committees were always understanding and friendly. The results were soon reflected in increased appropriations, which on my leaving were over $60 million a year more than I found them.

As soon as I could after assuming office, I began visiting the headquarters and the posts in the army. I not only established personal and friendly relations with officers and men; I also learned many needs and neglects that I could correct. I found the morale of the army low, and much needed to be done to improve living conditions, training, and confidence. I found families living in bachelor quarters with no place to cook. Some cooked in the bathroom. Others took the wives and children to the officers' mess at an expense they could not afford and where proper food for young children could not be obtained. I had kitchens added to bachelor quarters at a number of places and converted temporary barracks into comfortable apartments. I found wives at posts with tuberculosis and ordered their husbands to Denver to duty, where the wives could go to Fitzsimmons General Hospital.

The War Council, consisting of the secretary of war, the assistant secretary of war, the assistant secretary of war for air, and the chief of staff, met in the offices of the secretary at 1:00 P.M. daily, immediately after the cabinet meeting. Here, I presented all matters requiring approval or a decision by the secretary. The meetings were always harmonious, although the assistant secretary[9] was unfriendly to me from the beginning. He was, no doubt, influenced by the chiefs of supply branches over whom the law gave him complete control. The Joint Board, consisting of the chief of staff of the army, the deputy chief of staff, the chief of the War Plans Division, the chief of naval operations, the deputy chief of operations, and the chief of naval war plans, met weekly. They were charged with preparing, revising, and coordinating all war plans. The main war plans were designated as the Red plan for war with a European power, the Orange plan for war with Japan, and the Red-Orange plan for war with a European power and Japan. The European power especially considered was England, whose recent ten-year treaty with Japan and whose traditional unfriendly policies rendered her our most probable enemy. However, all countries were considered in Europe, Asia, and the Western Hemisphere.[10]

At the first meeting, the navy members submitted a war plan for the Pacific against Japan.[11] It was to be largely a naval war. The fleet was to assemble in Hanalei Roads in Hawaii with fifty thousand marines on transports and proceed to the Philippines, defeating the Japanese fleet en route. I asked if there was an estimate of the situation or a study on which the plan was based and was told that there was not. I knew Hanalei Roads to be little more than an open sea where the ships would be exposed to submarine attack, like the Russian fleet at Port Arthur in 1905.[12] I also knew from our studies of the Hawaii war plan that such a combined movement was impracticable and that it would require more than fifty thousand men to conquer the Philippines. Besides, a fleet could not protect transports while convoying such a large number of troop ships and supply boats. I immediately moved that the plan be referred to the joint planning committee of the Joint Board for study, preparation of an estimate of the situation, and a formal war plan.

The effort took a year and a half but produced a voluminous report in which every conceivable factor was fully presented. A sound war plan was the result. In general, the fleet was to be at sea based off the Pacific coast when war was expected. The fleet was to find and defeat the Japanese fleet and return to the Pacific coast, where transports were to be loaded with troops ready to sail. These would be escorted to the Philippines and relieve Corregidor. The troops stationed in the Philippines would retire to Corregidor, without resistance or losses, and hold the fortifications until relieved. Supplies for six months were accumulated. These did not include Filipino troops, whom no one could have considered as effective. The American troops and Philippine Scouts could hold Corregidor indefinitely.[13] No capital ships went to Hawaiian waters to stay, and no one could have thought of a capital ship, much less the fleet, entering Pearl Harbor. Ships in Pearl Harbor could not leave without exposure to submarine attack. A ship sunk in the channel would bottle up anything in the harbor for a long time. At that time, air attack from overseas seemed impracticable, but there were reasons without it to keep everything but a few submarines and small station boats out of Pearl Harbor.

The Red-Orange plan required the defense of both coasts and our northern and southern frontiers. We knew the capacity of England and Japan to mobilize and transport troops to Canada. Consequently, our mobilization plan required us to place a superior force along the Canadian border before we could be invaded. Similar plans were made for the southern border. These war plans were constantly updated, with the unanimous agreement of the Joint Board. My only difference with the navy was on land defense by naval aviation. The navy wanted Ford Island at Pearl Harbor and Rockwell Field at San Diego, California, for land-based planes to defend the land. I contended that the mission of the navy was to defeat the enemy's fleet and protect the sea-lanes, allowing the army to move troops and supplies. The mission of the army was to repel attacks on land. The principle of paramount interest was adopted as a rule. At sea, or where the shipping considerations were paramount, the navy would command. On land, or in questions governing land operations, the army would command. It is the only method for joint operations.

I found that all radio communications overseas and within forty miles of our coasts were controlled by the navy. I demanded the right of the army to operate radio for communicating with the army troops overseas and with transports at sea. At first, this was bitterly opposed by the navy, but at last it was adopted on the motion of a navy member and gave us twice the power that I had requested. The composition of the Joint Board during my incumbency was the ablest group of men I have ever known.[14] Subsequent events indicated the wisdom of all their decisions, plans, and policies.

I fully realized the potentialities of aviation and the need for increasing and improving the air corps. I cooperated wholeheartedly with Mr. Trubee Davison,[15] assistant secretary of war for air, and with General James E. Fechet,[16] chief of the air corps, in their efforts to increase the aircraft from eighteen to twenty-three hundred and to add to aviation personnel. In order to supply the additional air corps personnel needed, as no appropriations could be obtained for this purpose, other branches therefore had their numbers reduced. Money was secured for additional planes, but the losses were about as great as the increase, so progress was slow.[17]

Morale is necessary for efficiency, and officers must have reasonable comforts and recreation for themselves and their families. I found that the officers of the army and navy in Washington had little or no recreation because the country clubs would not admit them and they could not afford the fees. During the World War, some reserve officers had purchased some land and a roadhouse near Arlington, Virginia, and deeded it to the services for a country club. Nothing had been done to develop it. I effected an organization and obtained subscriptions to improve the house. I detailed an officer, Major Newman,[18] to build a golf course with prison labor, army transportation, and materials. They also built roads, swimming pools, and tennis courts. The old fort adjoining the house lot belonged to a former officer in my company. He authorized its use as a playground for young children while their mothers were playing golf or having entertainments. We also employed a nurse for the children and brought an excellent caterer from Fort Sam Houston, Texas, to operate a dining service. Dinner and lunch parties could be given at a small cost. The Fort Myer orchestra played for frequent dances. Initially, the navy would not give any help and would have nothing to do with the project. When it became a success, they stepped in and managed it. After I left Washington, the main purpose of the club was defeated by establishing a limited membership, preventing the very people for whom I intended it from joining.

To improve morale, the greatest needs of the army were housing and rations. There appeared to be no way to improve either. A limiting figure was placed on the budget, and the amount was too small for the needs under existing conditions for maintenance. Many troops lived in wartime cantonments that had deteriorated and were not suitable for habitation. Officers' families at many places were little better off. New airfields had practically no housing for personnel. On one of my inspections, the chamber of commerce at San Diego, California, gave me a large luncheon attended by city officials and army and navy officers. In his remarks, the mayor harshly criticized the army for the shameful way in which cavalry troops were living near there for the protection of the border. They occupied wartime barracks and were as well off as the men at many other places. In my reply, I stated that it was without precedent in history that a victorious army returned to its own country and was housed in conditions worse than its enemy prisoners during the war. This was a fact, as prisoners were well treated in World War I.

The next morning, as I was mounting my horse to review the troops at the Presidio of Monterey, California, a telegram was handed me from the secretary of war ordering me to report to the president[19] at once. Of course, I assumed that my remarks at San Diego had been reported and that I would be relieved as chief of staff. I had no regret and was willing to be sacrificed in such a cause. I canceled the rest of my Pacific Coast inspection and drove to Oakland, where I conferred with the corps area commander from San Francisco. In order to avoid reporters, I drove to Benicia Arsenal and sought refuge in the home of the commanding officer while waiting for the overland train. Before I reached Washington, the papers were full of the incident and vigorously defended me, demanding that conditions be improved. When I reached my office, I called the president's secretary, who told me to report to the White House the next morning. In the meantime, I prepared a statement of what I had said to give to the president. When I entered his office, he rose to meet me and shook hands cordially. Even his pet dog came and was very friendly. The president said: "Well, General, the papers seemed to have exaggerated what you said in San Diego." I told him just what I said, and he had my written statement. Nothing further was said on the subject, and we talked for several minutes in a friendly way. The Committee on Military Affairs of the House of Representatives immediately called for an estimate for army housing and soon appropriated over $20 million.

Up to the end of my tour, I secured over $90 million for army housing, with the president's approval. Different types of houses were designed for barracks and for officers' quarters in the different latitudes. I spent $1 million to repair the quarters in the Philippines and about as much for quarters in Panama, all of which had greatly deteriorated from termites and decay. Vacant permanent posts were occupied, and the old permanent buildings were renovated. The quartermaster general[20] opposed me and, off the record, endeavored to have officers' quarters made too small for comfort. But the committee approved the cost and type that I recommended. Some corps area commanders took little interest in using money allotted for repairs, and it was not properly used for maintenance of buildings. It was necessary to call in allotments and have requests from some commands sent to the War Department for approval.

Not long after the San Diego incident, the Senate Appropriations Committee informed me that the appropriation bill from the House carried $7 million for items not authorized by the House Committee on Military Affairs. The Senate Committee indicated that they intended to appropriate the amount for items that were authorized and asked me what I recommended. I told them that I wanted $4 million to make the daily army ration amount to fifty cents. It had been limited to thirty cents, on which men could not live. Moreover, the navy ration was fifty cents. The increase was immediately approved. I then asked for a $1.25 million to put the transport *Grant* in condition for the Manila run. All army transports, built as small freighters, were unsuitable for the long tropical voyage. This was also adopted. I had no trouble in using the rest of the amount.

The next morning, the secretary of war told me that the president wanted us to lunch with him. I replied that I knew why, and at one o'clock we went to the White House. The president soon appeared, and we three sat down to a table where the greatest abundance of delicious food was served. None of us had any appetite. The president was very angry and at once said: "General, what is this about increased appropriations for rations?" I then stated just what had happened. He replied: "You know you are breaking my budget, don't you?" I stated that the money was in the House budget; that I had requested no increase; and, that I was required to answer questions by the Senate Committee. As little more was said, the secretary and I left as soon as possible.

Again, I expected to be relieved. Almost immediately, the chairman of the House Committee on Military Affairs[21] called me and said that unless the president issued an executive order making the army ration fifty cents, the House and Senate would pass a bill fixing that amount. I had an executive order prepared and took it to the secretary of state,[22] explaining the situation. Nothing happened, but in a day or two the chairman of the Senate Committee on Military Affairs[23] telephoned me that this action would be taken by the Senate, thus depriving the president of his executive authority to fix the ration. When I reported this to the secretary of state, the order was signed. Then I issued an order for officers to supervise the messes and the purchase of articles of food. I found in my inspections that many took little interest in the feeding of the men, and I was compelled to take corrective action.

In order to be prepared for mobilization, I had the Supply Section (G-4) of the General Staff prepare a table of quantities and costs of all kinds of weapons, ammunition, supplies, clothing, etc. for a million men. This would be used as the unit for any number of millions. The table that I had prepared for the report of the Munitions Board in 1916 was a reliable estimate of the time for procurement. Thus, complete data would be available. The cost at that time for the table was about $1 billion. This was, of course, confidential. To my amazement, the table was published in a morning paper. When I reached the office, the secretary of war said that the president was very angry and thought that it had been published to alarm the country and force appropriations against his policy. He wanted to know who was responsible for the release of the information. I told him that I had no idea but that I was responsible for everything that took place in the staff and would assume that responsibility. I said that if the president wished to relieve me, I was quite willing to go. I assured him that I would try to locate the leak. The only departments who had the document were the G-4 Section and the Ordnance Department. I asked the chief of the Supply Section[24] of the General Staff to investigate and called the chief of ordnance[25] and made the same request. He showed great hostility and resentment against my asking him. I had been informed that he [the chief of ordnance] was one of the leaders in the cabal against me because of my termination of reappointment of chiefs of branches. His action revealed his animosity and was unbecoming of his status. I had no more to do with him. It was found that a reporter had told an officer in the G-4 Section that I had authorized release of the information. Instead of notifying the chief of the section or checking with me, the officer gave out the information. No harm could possibly have been done, but it showed the anxiety of the president to prevent any pressure to spend money. The table could have come into use only with the approach of hostilities, when secrecy would not have any effect. The president seemed satisfied, and I heard no more of it. The night after the publication, we had a dinner for the members of the House Military Affairs Committee. After dinner, one of the members asked me who was responsible for the publication. I told him that I was responsible for everything that took place in the War Department and assumed the responsibility. He said no more.

From that time, I had no better friend than the president. I concluded that he was not aware of conditions, and, instead of blaming me when he learned of them, he commended my action in trying to correct them. After our conversation, he sent for me from time to time. I would not know why he wanted me. He had me as the only weekend guest at his summer home in Wisconsin one summer, where he and his lovely wife treated me in the most informal and friendly way. We were reading the papers on Sunday morning when he said: "General, I do not like the proportion of American and Filipino troops on Corregidor, where there are 45 percent Americans and 55 percent Filipinos. I would like to remove any temptation for the Filipino troops to revolt." There was much unrest in the Philippines, but I had no idea why he gave it any consideration.[26] The Treaty of Washington in 1920[27] prohibited any change in the disposition of troops in the Pacific. I told this to the president and assured him that I would find a way to correct the situation. On returning to Washington, I ordered an antiaircraft regiment of American troops from Fort William McKinley to Corregidor, thus reversing the proportion. Nothing was ever said, although I expected Japan to object. Later, the president sent for me and said that his mother-in-law was very ill and might pass away at any moment. He wanted me to prepare plans for his trip to Vermont and to go with him to help meet the ordeal that it would mean. I canceled a trip to Panama and remained ready. I had the commanding general of the First Corps Area[28] prepare all details. His mother-in-law recovered but died after he was out of office, at which point the plans were executed, except that it was not necessary for me to go.

In one of our conversations, he told me that while he was spending the summer at Rapid City, North Dakota, he saw a cavalry regiment march past and the officers were all walking with the men. He thought that a colonel who would treat his officers the same way as the men must be a very efficient officer, and he wanted him detailed as an aide at the White House. This I did at once, but I did not tell him that such was the custom. The troops were marching from Fort Warren at Cheyenne, Wyoming, to occupy the vacant post of Fort Robinson, Nebraska, as we had no money for transportation. Once after a trip to Florida, the president sent for me and told me that he had visited the Bok Tower.[29] He was taken to the top to see the carillonneur play the chimes. He was amused to see the man in a bathing suit jumping and stamping on the pedals while he struck other bells with his hands. He had a lively sense of humor.

At the request of the House Committee on Military Affairs, the War Department prepared a study on promotion and rank in the army. Among its recommendations was to give the chief of staff of the army the rank of general to correspond with the rank of admiral, which was the rank for the chief of naval operations. In transmitting the report to the president, I prepared an endorsement for approval, except that the promotion should not apply to the present incumbent as chief of staff. I did not want my enemies to say that I had manipulated any promotion for myself. In forwarding it to Congress, the president stated that he approved it only on condition that it did apply to the present incumbent. It passed Congress accordingly. The navy fought it bitterly but met with no sympathy in either house of Congress.

At an army relief party at Washington Barracks, the president was amazed by a display of the different army rations in our history. He felt that it was intended to reflect on his original attitude. My dear wife and I were included in most White House receptions and in several dinners. When he left the White House, we were invited to a special dinner in our honor. The highest form of courage is to see one's errors and to correct them without resentment. This was a quality of Mr. Coolidge, found in few men. I had experienced it but rarely. He was the last real president who conformed to his duties under the Constitution. His high courage, clear understanding of the national economy, and fidelity to duty were without a superior. In his semiannual budget meetings, attended by the cabinet, the heads of the army and navy, and the public, he educated the public on the need for economy in government. He said that one necessity for national defense was being out of debt so that the resources of the nation would be available for war. He approved of the liberal detail of officers in Washington, where they could learn the administration of the War Department and of the government.

As a result of his recommendation, on February 23, 1929, I was made a general. When my nomination was before the Senate, my enemies endeavored to influence votes against me. While at that time confirmations were in closed sessions, I was informed that three senators, all from the South, voted against me. Yet, I was the only officer from the South ever to be made a full general.[30]

Ever since the world war, I had strenuously advocated an all-purpose field gun, preferably the 105-millimeter howitzer, to attack both land and air targets in accordance with the mission of the field artillery. Both the Ordnance Department and the field artillery were strongly opposed to it. When a vacancy occurred as chief of field artillery, I had Major General Fred T. Austin[31] appointed, an officer who agreed with me and to whom I gave the mission of trying to influence the field artillery. It was useless for me to approach the chief of ordnance[32] on the subject. The commanding officer of the Watertown Arsenal, Massachusetts,[33] was an old friend, and he adopted my views. He used two 75-millimeter guns on tripod mounts for firing at planes or ground targets with all-around traverse. The guns could be lowered from wheels or mounted for mobility in less time than the 75-millimeter gun could be prepared for action. When the guns were put through the most severe tests at the proving ground, they performed satisfactorily. I went to the proving ground to see the test, but the officer in charge was very discourteous, no doubt influenced by the chief of ordnance, so I left. Nothing was done as to their adoption, and, when I left office, the project was abandoned. Such a gun would have been invaluable in World War II, when the field artillery was forced to use the 90-millimeter antiaircraft gun against ground targets. The 105-millimeter howitzer was adopted after the war began, but it was adapted only to ground targets. I also tried to expedite the manufacture of a semiautomatic infantry rifle, but the Ordnance Department, which had complete control of it, again procrastinated over trifles.

I realized that tanks or armor would play a decisive part in the next war. My conception of battle was to have a large force of artillery and aviation neutralize an enemy assault by heavy shelling and bombing. A large force of tanks in successive lines and with suitable intervals between tanks would then dash across the enemy's positions. These would be followed by motor-borne machine guns that would build up a line of intense fire against any enemy resistance. The infantry would then follow and exploit and occupy the position. I also greatly increased the number of machine guns with the infantry, as I had demonstrated in Hawaii that one machine gun was equal to a platoon of infantry in gaining superiority of fire. I tried to demonstrate this form of attack at Fort Meade, Maryland, and at Fort Benning but found officers unwilling to depart from the old destructive infantry assault.

No new material had been developed since the war. I requested the chief of ordnance to let me know what had been done to develop tanks. He sent a truckload of files on old model tanks to my office. I sent them back because actually nothing concrete had been done. I told the Supply Section (G-4) that I wanted designs for a light, medium, and heavy tank without limiting their weight. I wanted efficiency whatever the weight. Mr. J. Walter Christie had been a successful designer of tanks, but the Ordnance Department would not deal with him.[34] It evidently wanted no design but its own. One day, drawings for a light tank were sent to my office with a request that it be standardized for manufacture. On inquiry, I was told that the tank had not even been manufactured, much less tested for acceptance. Of course, I sent the drawings back not approved. On another occasion, I was asked to witness the performance of a light tank built by the Ordnance Department at Fort Meade. It was impossible from every point of view. It was manifest that I could not look for any new materiel.

The neglect of the Ordnance Department, which was charged with design and procurement, and the impossibility of compelling it to act were due to the vicious provision of the National Defense Act that excluded it from control by the chief of staff and had it operate under a civilian assistant secretary of war, who could not know what was required. I considered it necessary to organize a mechanized force for development of tactics, organization, and equipment. Accordingly, I ordered a battalion of infantry, a battalion of field artillery, and suitable proportions of cavalry, engineers, and the services to Fort Eustis, Virginia, and equipped them with world war material, tanks, reconnaissance cars, etc. I selected a very able officer[35] to command. The personnel were sent to the quartermaster motor repair shops to learn to care for and operate the materiel. This was the origin of what in World War II became the armored force. Great progress was made, and eventually the command was transferred to Fort Knox, Kentucky. If the Ordnance Department had been equal to its duties and had cooperated, or if the chief of staff had had authority over the Ordnance Department, the army would have been much better prepared for World War II.

The Organized Reserves, the Reserve Officers Training Corps, and the CMTCs, which were all administered by the adjutant general's office, had no adequate system for training, promotion, etc. The reserves were dissatisfied with the lack of a promotion system, and young men were rapidly advanced over older, more experienced officers. Training was unsatisfactory. I set up an office for all civilian components as a part of my office and supplied it with a chief and a complete staff. I had adopted a system of promotion for the reserves by which an officer would serve three years as a second lieutenant, four years as a first lieutenant, five years as a captain, six years as a major, and seven years as a lieutenant colonel, thus attaining a colonelcy in twenty-five years. I also required attendance at local instruction and training camps. The result was high morale and improved efficiency.

Students and faculty at many colleges objected to the ROTC. Pacifism was rampant among students at the time. I placed the ROTC under corps area commanders and instructed them to visit the colleges and try to win the friendship of college presidents. I also visited a number myself. The results showed a great change for the better. I removed a few units from especially unfavorable places. The CMTCs improved and were well attended. The chief of the Civilian Components Section in my office spent much time traveling and holding conferences.[36] We selected a high type of officer for duty with all the civilian components. Although every effort was made, increased appropriations could not be obtained to support these activities. The national guard divisions generally had summer training camps. I was able to assemble the Thirtieth Division from some Southern states at Camp Jackson, South Carolina, by combined marching and trucking the troops.

I endeavored to increase the national guard from 190,000 to 205,000 but succeeded in obtaining money for only 145,000. The Coolidge economy measures continued after his leaving office.[37] I found that the regular troops had no marches or field training and no money for maneuvers. I ordered that annually all troops should have at least one week's marching and camping. In one case, when I found that a lieutenant colonel and a major at a post sent the battalion under the senior captain for the march, I ordered both to join the troops. On my visit to another post, the commanding officer was showing the effects of drink when he met me. I relieved him on returning to Washington.

The U.S. Military Academy had long been administered by the adjutant general. As a school, it belonged under the Organization and Training Division (G-3) of the General Staff, so I transferred it accordingly. The defense of its budget was a part of my duty. I was able to increase appropriations to build the mess hall and a large number of officers' quarters. I selected able superintendents and enabled them to improve academic departments. The system of substantiating examinations of candidates was established to prevent ignorant men with high school certificates from entering and causing useless expense and vacancies in the Corps of Cadets.

In another reform, I succeeded in rehabilitating many military prisoners by having them serve their sentences at their posts instead of at military prisons. They thus saved the troops at the posts from a large amount of fatigue duty. On one trip to New York, I had noted serious conditions at the Brooklyn army base. Men arriving from overseas for discharge were moved on open barges to Fort Hamilton in winter, exposing them to severe cold in tropical dress. There, they were required to perform fatigue work until discharged. As a result, few reenlisted. Men are most sensitive to impressions when they enlist and when they are discharged. I had prisoners sent to the base to do all the fatigue and warehouse work. Barrack rooms were fitted up at the base for overseas casuals,[38] who were well fed and did nothing but make their beds. The result was that they went out feeling well treated and many reenlisted.

There was no place at the Brooklyn base or San Francisco for men and their families while waiting to embark for overseas tours, and most could not afford the price of hotels. I had quarters prepared for officers and men and their families, where they could stay comfortably and obtain meals at the cafeteria at reasonable cost. I did the same at Fort Mason, California. On one of my visits to the Brooklyn base, I noted a casket standing on end. I had a small room prepared as a chapel where caskets were kept properly until removed. When Mr. Hoover was inaugurated as president, March 4, 1929, I was grand marshal of the parade and attended the inaugural exercises at the Capitol, afterward standing by the president as the parade passed the White House. There was a small reception in the White House afterward.

The new secretary of war was Mr. James W. Good,[39] whom I first met that day. The day after the inauguration, the president sent for me and read dispatches about the Mexican revolution that had just broken out. He asked what we should do. I told him that we should be strictly neutral and treat both sides alike. We should also warn American citizens not to expose themselves to danger on the American side of the Rio Grande, as the government would do nothing to fix responsibility or exact reparations. He agreed, and I issued orders accordingly. Our garrisons along the border remained friendly and received and cared for refugees on both sides. The result was goodwill toward us when the insurrection was over. There were no incidents involving civilians, as in former revolutions. When Mr. Good took office, the staff gave him and the retiring secretary, Mr. Dwight Davis, a lunch, at which I presided. In presenting Mr. Good, I referred to his having established the budget system in Congress and jestingly warned him that he would now see it operated. In his remarks, he said that one must be governed by the circumstances in applying the system.

The president told the new secretary of war to have all officials in the War Department submit their resignations at once. I did so, although I had never heard of such a procedure for military men. A day or two later, a gentleman came to my office and said that his name was Patrick Hurley.[40] He had just been to the White House, where the president asked him if he would like to be assistant secretary of war. He replied: "Mr. President, I would crawl on my hands and knees to the War Department to be assistant secretary of war." The president told him that he would appoint him but could not without the consent of the vice president, Mr. Charles E. Curtis.[41] Mr. Hurley had heard that I was a close friend of the vice president and wanted to ask me to get his approval. I called the vice president and went to his office at the Capitol. When I told him what I wanted, his face clouded, but he said: "General, I will do this for you." He then called the president and gave his consent. Mr. Hurley at once became assistant secretary of war and was always most friendly to me.

Not long afterward, he asked me to find a suitable officer to be his executive.[42] I recommended a very able officer who for years had written me the most devoted and loyal letters. I needed friends in that office, which controlled all of the supply branches. It was not long before I learned that this officer was working against me and trying to turn the secretary against me. He had been friendly to me only with the hope of gaining my confidence and using it for his own advantage. He ended his career in revolt and narrowly escaped court-martial.

President Hoover sent a directive to the War Department for a study on the reorganization of the government. I had this made with a view to placing all agencies in war under a cabinet-level department in peacetime. Some would contract in peace and expand in war, and vice versa, but every agency would be prepared to function efficiently when war came. This was included in my final report. Many years later, Mr. Hoover headed an executive board to make the same study.[43] My report received no attention but would have been invaluable during World War II if it had been adopted.

The defects of the law exempting the chiefs of branches from control by the chief of staff would largely have disappeared if the military head of the War Department could control the warehousing and issuing of supplies. I found that it would take six months to distribute the medical supplies from depots where they were concentrated. Transportation, clothing, etc. were similarly concentrated. I wrote the assistant secretary of war a letter asking that procurement, which, under the law, was a function of the supply services, should cease on acceptance. Thereafter, the warehousing, distribution, and issue would be under the chief of staff. Supply is an essential function of command, and I wanted to distribute all available stores according to where they would be needed for mobilization. The secretary never would act on my request, although I felt that he agreed with me. The influence of the chiefs of branches was too great to resist. I discussed the matter with the secretary, and he agreed with me but thought that action should be deferred until we had completed the change in chiefs of branches under the policy of not reappointing them.

Mr. Hoover had a camp on the Rappahannock River. He invited me to spend a weekend there with the cabinet to discuss the budget. I rode with him and the secretary of war and the secretary of the interior.[44] When we arrived, we put on working clothes and rubber boots and worked for some hours building a dam over a stream. On Sunday morning, I was asked to explain the army budget to the president and the cabinet. All seemed satisfied, and Mr. Hurley complimented me heartily. I admired Mr. Good, the secretary of war, greatly, and we became warm friends. One day, he told me that the president had given him a specially built Cadillac car and he wanted me to have it. I suggested that the assistant secretaries might want it, but he said, no, he wanted me to use it. Within a brief period, Mr. Good underwent a serious operation and died.[45] There were no appropriations for his funeral expenses. A fund of about $700,000 was subject to control of the War Department from the liquidation of canteens in the World War. I used this for all expenses, including a suitable casket and a special train to take the casket and the family and the cabinet and the committees of Congress to his home for burial. Of course, I accompanied the party and the family. His death was an incalculable loss to me and to the War Department. He was a noble man, a lofty character, a wise administrator, and a loyal friend. His experience as a lawyer and a political leader eminently qualified him for his office, and, had he lived, many necessary reforms would have been accomplished for the good of the army. Mr. Hurley became secretary of war. He told me to find an assistant secretary of war, implying that I might find someone who would help me in the difficult situation that existed. I requested Mr. Crane,[46] a prominent manufacturer from Massachusetts whom I did not know personally but whose reputation eminently fitted him for the office, to come to see me. We discussed the matter at great length, but he finally declined.

I then called Mr. Gilchrist Stockton[47] of Jacksonville, Florida, a Rhodes scholar and an able man, to come to Washington. He at once accepted eagerly. When he went to the White House to tell the president, whom he knew well, the president at once told him to go to Austria as minister as there was some difficulty there. He could not persuade the president to let him be assistant secretary of war. I then requested another industrialist in Massachusetts[48] whom I did not know personally to take the office, and he accepted. However, I was not able to secure any concessions from him after he came under the influence of the chiefs of branches.

In my final report as chief of staff, dated November 20, 1930, I made a complete exposition of the law, the practice, and the evil effects of the nonmilitary control of supply and presented a recommendation to place the supply situation on a sound basis. From the beginning of my duty as chief of staff, I placed mobilization as of the first importance. This is fully discussed in my final report. In principle, the tactical and supply units required for six field armies were decentralized for mobilization to the corps area commanders. The old idea of having mobilization progress only as troops could be supplied and armed was discarded, with plans prepared to expedite supplying and arming. The situation under the Red-Orange plan would not permit delay to construct cantonments but must proceed more rapidly than the enemy could land troops on territory adjacent to the United States. Housing was to be found in hotels and other available buildings. Training would be conducted in parks and areas adjacent to cities. Public hospitals would be used as far as necessary for the sick and injured. A draft act was prepared and ready to submit to Congress, which would have passed it in thirty days. Draft boards were selected, and we believed recruits would be pouring in within sixty days. With proper housing, food, and care already available, the health of recruits would be preserved, and they would be gradually hardened by training. Practice mobilizations on paper were carried out, and the operation was found to be successful. An officer from the Organization and Training Division (G-3) of the General Staff visited all corps area headquarters and explained to the commanders and staffs all the details. Each study was given to the roles of the different arms in the next war and to the development of aviation and armor. This subject was fully discussed in my final report.[49] Later, the experience of World War II fully supported the conclusions and the policies that I advocated. General Staff control over the supply services was one of the first changes forced on the War Department. The long period of mobilization might well have lost the war. Fortunately, the enemy permitted a delay, which we could not assume in advance. My final report must form a part of any account of my tour of duty as chief of staff. It was prepared by unusually able officers with whom I had constant conferences. We were unanimous in the conclusions reached. It contained a discussion of the organization of the executive department of the government that will always be sound and necessary for war and peace.

The secretary of war asked me to recommend a list of promotions for general officers. When I did so, he said he would carry it out. It was necessary to give precedence to older colonels and brigadier generals, or they would retire before promotion. The president would not promote an officer within one year of retirement, as was done in the navy and had long been done in the army. My recommendations were carried out after I left office.

Although it would not be practicable to record the many accomplishments that were routine but had marked influence on the army, I was proud of a number of reforms that were completed. The restoration of the blue uniform contributed to pride, self-respect, and morale. A soldier's conduct reflects his dress, and a poorly dressed man cannot be a good soldier.

A great many officers who had graduated from the Command and General Staff School were not recommended for the General Staff because they did not receive a high enough grade. They felt that they were unjustly treated. A board of general officers was appointed to examine their records and recommend all who were efficient for the General Staff. Many capable officers were thus made available, and their morale was restored. Officers at the Command and General Staff School were rigidly marked, and, unless an officer had a very high grade, he was not recommended for the General Staff. Not only the officers but also their wives were made to feel inferior, with even social distinctions made. Not long before, an officer at the school had committed suicide, presumably on account of his marks.[50] I had the school stop all marking. If an officer was satisfactory, as practically all were, he graduated equal to the others. If he was not satisfactory, he was to be relieved. The effect was highly satisfactory.

Another change involved cases of spinal meningitis in camps and on transports and, at times, men killed by plane accidents who were so mangled that special preparation for burial was necessary. There were no funds for such expenses, and I used the residual world war post exchange funds to purchase metal caskets and provide special services of undertakers.

The office work required long hours. Many visitors called on important subjects. I seldom went out to lunch but had a glass of milk and a sandwich brought to the office. I remained after office hours until time to go home and prepare for dinner. The two faithful secretaries insisted on staying until I left, though I did not often need them. There were many inspection reports, Class B proceedings,[51] staff recommendations, court-martial proceedings, and reports of boards that required careful consideration. Numerous letters from many sources were answered or referred to departments. I tried to have all letters couched in courteous language. Congressmen often wanted what could not be done, and explanations were made. For example, a senator wanted the body of a dismissed officer interred in Arlington National Cemetery. This could not be done. But I found that he had a superior enlisted record, so I ordered him interred as a soldier. A mother told me that her paralyzed child would suffer on an army transport ship without his dog. I ordered the dog to be taken on the transport.

Throughout my four years as chief of staff, we did our full part socially. We generally went out to dinners or entertained. My dear wife gave dinners as beautiful and abundant as any in Washington. We entertained the cabinet, diplomats, congressmen, senators, army and navy officers, and their wives. We usually had eighteen guests. The seating was arranged by the protocol division of the State Department. Once each month on important occasions we gave receptions that were well attended and at which the guests remained long. The grace, charm, and cordiality of my dear wife captivated all who saw her. She had the wives of other officers assist her, and no caller was neglected. Her entertaining went far to make friends for the army among important people. In our many years of experience in Washington, and from reports after we left, I believe that the army was never as well represented or did its part as completely as when my dear wife presided over the residence and the social program of the chief of staff. It cost all of our pay and allowances and much of my dear wife's legacies from relatives, but we felt rewarded in performing our obligations and in the good that came to the army.

General Austin, the chief of field artillery, of his own accord transferred our son to a battery at Fort Myer; consequently, we had the happiness of having him and Julia near us. They occupied half of a single set of quarters because of the crowded condition of the post. An event of equal importance to their marriage was the birth on September 9, 1929, of their twin children, Charles Pelot III and Julia Reeder. We went with Charles and Julia to Walter Reed Hospital late at night. Charles drove very fast and barely reached the hospital in time. He, his mother, and I sat in the car and waited. Soon, a nurse came out and told Charles that he had a son. In a very few minutes, she returned and said he also had a daughter. Judy was a great surprise as the doctors had not detected her. She was very small and was placed in an incubator and fed with a dropper for several days. I always called her a miracle child for her life was very feeble. Julia and the babies and, of course, Charles stayed at our house for several weeks. An old mammy who had been in my dear wife's family took care of them. They were called Punch and Judy, and the names stayed with them. They were healthy and grew normally. "Mammy" said they were very smart, and she made no mistake. After they returned to their small apartment, I stopped by each morning on my way to the office to see them. They have been a great blessing in our lives and are all we could want them to be. They were christened in our house.

At the end of my tour, the president gave my dear wife and me a small dinner. After dinner, in his study, he asked me if I would like a place on the Federal Communications Commission. I told him that I would but heard no more about it. The Committee on Military Affairs of the House of Representatives adopted a resolution expressing commendation and appreciation of my services. Before I left, my friends came to the office to take leave. I was far more satisfied to have done my duty and made enemies than I would have been if I had failed in my duty and retained their friendship. I have often said that I wanted my epitaph to say that I never failed to perform my duty for fear of what others would do. Like Byron's Childe Harold, my motto has always been: "Here's a sigh for those who love me, and a smile for those who hate; and, whatever sky's above me, here's a heart for every fate."

After the War Council on April 20, 1930, I left a clean desk and ended my active career. On returning to Fort Myer, a battery of field artillery under command of my son for the occasion fired a salute of seventeen guns, and my dear wife and I drove to Union Station. Here a few friends said good-bye to us.

Chapter 28

The Wanderer's Return

In reaching Jacksonville the next morning, Senator Fletcher[1] and a few friends met the train. We reached our home in Eustis, Florida, in the afternoon. A comfortable boardinghouse was available, and we stayed there while the home was being repaired.

Sometime before the expiration of my tour as chief of staff, a financial corporation in Boston had asked me to join it on retirement. I wanted to do so, but the Depression had so reduced the volume of business that the idea was abandoned. On reaching Florida, I received many letters asking me to run for governor or U.S. senator. I considered the possibilities. Much money would have been required, and we had only some small legacies left to my dear wife. They would not have been enough, and I would not have used them. A man offered me $25,000 for a campaign to be governor, but I declined on the ground that he would control me. My dear wife also objected to a political career. I was also offered the position of road commissioner in a Southern state at a large salary. It was involved in politics, and my dear wife strongly advised against it. But, when a telegram arrived that offered me the presidency of the Citadel,[2] the military college of South Carolina, she thought well of this, and, after visiting the college, I accepted.

Some months later, I was invited to address the joint session of the South Carolina general assembly. In the meantime, the city of Tampa gave us a large reception, and the Florida legislature invited me to address a joint session. We stayed with the governor while in Tallahassee. We also drove to various places in the state at some of which I was invited to speak. Our winter in Florida was one of the happiest periods of our lives. While we had been in Washington, my wife bought a Chevrolet for her personal use, as the official car should not be used for private purposes. She learned to drive, but a driver in Washington took the car to Eustis. I learned to drive, and we thus became independent. The people in Eustis and the nearby towns were most friendly, and we saw much of them.

My retirement was effective March 31, 1931. In June, we drove to Charleston for the commencement at the Citadel. Colonel Bond,[3] the retiring president, and Mrs. Bond gave us a reception. We then drove to Washington for a visit to our son's family at Fort Myer and returned to Vella Crucis, North Carolina, where we spent the summer. My dear wife was not well, and, when we returned to Charleston on September 1 to take up my duties at the Citadel, she became quite ill. I took her to Walter Reed Army Hospital, where she remained for some weeks until she recovered. I returned to Charleston and entered on my duties as president of the Citadel on September 12, 1931.

Chapter 29

The Citadel

On my visit to the Citadel in January 1931, I asked the chairman of the Board of Visitors[1] if my decisions on matters of discipline would be final, and he assured me that they would be. I would not have gone otherwise. When I accepted the appointment as president, I requested the War Department to change the orders of an officer designated as professor of military science and tactics and to detail an officer who had been my public relations officer in the War Department, who was eminently qualified for the position, and who was also qualified to be commandant of cadets.[2]

I had noted that the buildings were in bad condition. There were no paved roads on the campus. The furniture in the cadets' rooms was obsolete, and the beds had no springs. The two boilers of the central heating plant were inadequate and were hand fired. There were only six apartments on the campus for officers, the remainder of the faculty and staff being housed in the Old Citadel, which was in bad repair. Alumni Hall was used for indoor games, dances, and chapel services, with the cadets sitting on bleachers. The finances were insolvent, and old bills could not be paid. The kitchen was in bad condition. Bread, pastries, etc. were purchased. The enrollment had been falling off greatly. Only a few of the faculty had degrees above that of bachelor, and no effort had been made by members to improve their degrees. There was no system of promotion of officers, and rank was established in the lower grades. The discipline of the cadets was poor. There was much drinking in barracks, and women were introduced during weekends for immoral purposes. The budget had been prepared for the next fiscal year without providing for a large deficit. The library, cramped in a barrack, consisted largely of old books and reports, most of which were of little use. The laundry had worn-out wooden washers and an aged mangle, and all pressing was done by hand. The only transportation was a very lame mule, which died in a few months, and a wobbly cart that fell to pieces. The grass on the campus was cut only in the fall of the year for hay, and this was done by a man who took half for himself. There was no storehouse and, of course, little to store. Trunks, arms, property, the cadet store, the canteen, the post office, the commandant's office, and several classrooms were in barracks.

The new commandant of cadets at once began to improve discipline, but progress was slow. I took measures to reduce expenditures by requiring that no money could be spent without a requisition approved by me. The publicity of my appointment had increased the enrollment by probably a hundred over what it would have been. Very soon, I appeared before the committee of the general assembly to defend the budget. In the senate finance committee, I stated that a considerable deficit existed and that money was needed to eliminate it. A member immediately took me to task very harshly for a deficit for which I was in no way responsible. I told him that I had found it and that, unless something could be done to meet the obligations, the college might close. I then told him that I was a general in the army and was entitled to respect, that I could not submit to such treatment as he had shown me, and that I would resign immediately. I then left the room. On returning to Charleston, I submitted my resignation to be effective at once. The incident was carried by the papers all over the country and in some of the foreign press. I had many letters commending me. A few papers and letters criticized me. One letter, not signed, must have come from one of my father's old enemies in Florida. Immediately, the Corps of Cadets signed a petition for me to stay. The Board of Visitors met and requested me to withdraw the resignation. I told them that I would do so because of the petition of the cadets. After that, I was always treated with respect by the committees in the general assembly. The appropriation was not increased to pay the deficit. I, therefore, reduced all salaries and wages, discharged linenearly all laborers, and made practically no purchases except for food. By the end of the term, all bills were paid, and a slight balance was distributed as salaries and wages. This system was continued for the following year, when more was done for the upkeep of the property.

The college adopted a publicity campaign to procure more students, without which it would have failed. My dear friend Colonel Clark Williams, of New York and Camden, South Carolina, at once gave me five full paid scholarships. I called on the branch societies of the First Division to advertise the scholarships and select candidates. This secured press notices in all parts of the country. The

Chicago Tribune, owned and published by my close friend Colonel Robert R. McCormick, gave us a great deal of space and made the Citadel known throughout the prosperous Middle West. The result was a rapid and constant increase of students. Colonel McCormick then established two full paid scholarships from Illinois, and Colonel Williams increased the number of his scholarships, thus making the college better known. Colonel McCormick also gave full-page publicity of the Citadel in the *Tribune*.[3] Other scholarships followed by Senator and Mrs. Metcalf[4] of Rhode Island, Mrs. Frank G. Geary[5] of New York, and Colonel and Mrs. D. P. Quinlan,[6] who were close friends. Colonel Williams also donated a projector for motion-picture instruction.

Not long after I had become president, an incident happened that forced me to a grave decision. It was found that a cadet had scratched out the name of another cadet and written his own name on some clothing. This was stealing, and, under the college regulations, I suspended him for dismissal. He appealed to the Board of Visitors, which acquitted him. The commandant of cadets at once resigned. I was confronted with resigning or accepting the reversal of the board in all future cases of discipline. Ordinarily, the failure of the board to support me in carrying out its own regulations would have discredited me and so weakened my authority as to render my efforts useless. I finally decided to take the position that I had done my duty and that the exercise of the superior authority of the board could in no way reflect on me. I adopted this policy, and, although the board constantly reversed my action on appeals to it against all classes of punishment, I felt no concern as to the consequences. The cadets knew that I had done my duty and that I was not influenced by any action that the board might take.[7]

With the increased attendance, I was able to increase salaries and wages for faculty and staff and to improve living conditions. I replaced the old iron slat bunks with spring cots. The rotten plaster in bathrooms and toilets in barracks was replaced by tiled walls and floors. Mechanical stokers and Bailey control meters were installed in the boilers, and a new boiler was purchased. The school built a bakery and installed an oven for bread and pastries. Machinery was purchased for the laundry. A program for buying books for the library was adopted. Repairs and painting were gracefully accomplished. We purchased a secondhand truck, a secondhand tractor, and a mowing machine. Ultimately, a complete transportation supply was procured.

In the early days of the New Deal under President Franklin Roosevelt, we obtained some Work Projects Administration labor and built tables and book and magazine racks for the library, two additions to the laundry, a walk of one mile to deep water in the Ashley River for boating, and an outdoor swimming pool with temporary dressing rooms, showers, and toilets. When the federal government began to make grants of 45 percent for the cost of buildings, a construction program was prepared, and the general assembly authorized the college to issue bonds for the remaining 55 percent of the cost of projects. These bonds were secured only by the right of the holders to operate the college in case of failure to amortize the bonds. Yet the government sold them to insurance companies at a large premium. The first group of buildings to be constructed consisted of a chapel, with seating for fourteen hundred, a mess hall with seating for twelve hundred, a kitchen, and twenty-four faculty quarters. These were followed by an engineering department building, a ward to the hospital, two additional mess halls seating six hundred each, an armory, a swimming pool, an activities building with an auditorium, and sixteen family quarters. Additional classrooms and a library were added to Bond Hall, nearly doubling the size of the building, including three large chemistry laboratories, physics laboratories, and special business administration rooms. A powerful testing machine and laboratories were furnished to the engineering department. A new barracks was constructed. All roads were paved. The quartermaster's and commandant's offices, the post office, the armory, and classrooms were removed from barracks, and all the resulting space was converted into rooms for students. A large storehouse was constructed for trunks and property. The best form of lights was installed in all barracks, lockers for clothing were substituted for clothes presses in all new barracks, and better furniture was provided. The state then appropriated money for a second new barracks. The entire underground drainage and sewer system and steam pipes from the central heating plant were reconstructed. The athletic field, with running tracks and four additional tennis courts, was built, gymnasium equipment and boats were purchased, and projectors and an organ for the auditorium were purchased. Additional officers' apartments were made in the Old Citadel, and all apartments were improved and the roof of the building repaired. When World War II came, the college had nearly nineteen hundred cadets and up-to-date buildings and equipment.

After World War II, another building program was initiated. With $600,000 appropriated by the state and $400,000 saved during the war from government rentals and tuition, there were added an academic building with sixty-four rooms and an officers' apartment building with sixteen apartments. When I left, appropriations of nearly $0.5 million were available for a new laundry plant and a boiler plant to which another new unit had previously been added. There was also remaining $160,000 from the Citadel Reserve Fund. The legislature had just authorized loans from the state amortized by tuition fees. This would enable the Citadel to borrow at least $2 million. With this, I planned a building program as follows: an apartment building for twenty-four families, an engineering building, a hospital ward, an activities building, a complete armory by adding offices in the front and an arms room and gymnasium in the rear, and the completion of Alumni Hall. Donated funds and available money provided for erecting a carillon and tower adjoining the chapel. I also planned another barracks between the activities building and the north gate. In the meantime, ample and comfortable houses had been built for key civilian employees and storehouses for every need.

When I went to the Citadel, the college had no standing in the educational world. Its degrees would not admit graduates to full standing for postgraduate courses. The University of Illinois sent me a rating of the Citadel as a third-class college. Within a few years, I applied to the Association of American Universities to have the Citadel placed on its approved list and to the Committee for Engineering Education for the civil engineering department to be placed on its approved list. Dr. Bowles,[8] the registrar of Columbia University, was sent to inspect the college. When he was leaving after two days, he asked me what I thought I had done most for the college. I told him getting it out of debt and operating on a greatly insufficient budget. He said, no, that I had given it a faculty. The college was at once placed on the approved list and had the same standing as any other college or university. The committee sent to inspect the civil engineering department took the position that a military college could not train engineers because of the time devoted to military training. I prepared a document showing that all engineering education in the country stemmed from West Point. In the early nineteenth century, it was not taught anywhere else. The textbooks were in French because there were no English books on engineering. The early railroads, international boundaries, coastal surveys, and all of the early instruction in engineering were provided by graduates of West Point. The Corps of Engineers is still charged with handling all public works of far greater magnitude than any civil undertakings, including the Panama Canal. The argument was neunassailable, and the engineering department was placed on the approved list, thus becoming the first of any institution in the state to have this honor.

As fast as funds could be obtained, we added a department of classics to teach Latin, Greek, and philosophy for the benefit of theological students, a department of electrical engineering, and a department of political science and government. The curriculum for the last was prepared by the State Department as I wanted our graduates to go into the Foreign Service. When I received an honorary degree from Brown University, I fell in with Mr. Cordell Hull,[9] the secretary of state, whom I had long known. We marched together in the procession and sat alone at a lunch table. He talked to me very freely of his problems. I explained my plan for a department of political science and government and told him that I wanted the State Department to advise me as to a curriculum. He told me that, if I sent a professor to the State Department, he would have the proper section cooperate. This was done. In this department, I have insisted that the relations between government and armament be stressed. I could not find that subject in any of the curricula of over forty colleges where the department exists. The education of men who become responsible for national policies was void of the most vital problems of peace and war.

When I came to the Citadel, hazing was practiced to an extreme degree. At one inspection, Colonel Lang found that over thirty fourth classmen showed bruises from being beaten. All first classmen had fourth classmen detailed to perform all menial services in caring for rooms, clothing, arms, etc. In some cases, they were required to furnish cigarettes and other articles to upper classmen. For the first few years, I inspected barracks on Saturday mornings and awarded merits to rooms in excellent condition. I then found that the preparation of many of these rooms had been done, not by the occupants, but by fourth classmen. Freshman regulations were prepared and approved with a view to eliminating this evil. These were of little avail. The situation culminated in a mutiny of the first class when Colonel McMurray,[10] the commandant of cadets, canceled the freshman regulations. I suspended the leaders, but the Board of Visitors acquitted them. Since then, hazing has been greatly ameliorated, but it can never be eliminated until cadets who haze are dismissed.

The practice of bringing women inside the barracks for immoral purposes persisted to the extent that on at least one occasion late at night a woman arrived in front of the barracks in a taxi and left it saying that she was going to the room of the officer of the day. It is believed that now this evil has been eliminated. It also was common for cadets and girls to drink at hops and then to leave the hop room and sit in cars. This was stopped by requiring all who left the hop room to leave the campus.

There had been no receiving line at the hops, and all came and left without any formality. A receiving line was established. Cadets then decided that the cadet who introduced the guests should have his date stand by him. This was very acceptable. It was not unusual for cadets to appear at dances under the influence of liquor, although for several years previous there had been practically no drinking at dances. At first, cadets did not wear gloves to the hops and were averse to doing so. When our athletic teams went to West Point, I had them take gloves for the hop. When they returned, they reported that the West Point custom was to wear gloves, and the cadets gladly adopted that practice.

With the construction of the activities building, a hostess and reception room were provided. Mrs. Jesse Gaston was employed as hostess.[11] She conducted dancing classes and company dances during cadets' free time, and cadets have thus met many nice girls in Charleston. The room was also frequented by cadets and veteran students for games and relaxation. The canteen was commodious and patronized all day. Cadets and veteran students met in it, and it has a social influence. The poolroom was also open all day, and cadets of all classes mingled in it. Post office boxes gave each person privacy and accessibility to mail. A well-conducted barbershop, a cleaning plant, and a printing office with the best equipment also operated efficiently.

When the chapel was constructed, a schedule for services was established. At first, the Catholic cadets were not allowed to go to it. On Sundays, the Episcopal cadets had communion at 7:00 A.M., the Catholic cadets had mass at 8:00 A.M., and the Protestant cadets had services at 9:00 A.M. The ceremony of marching to the chapel with the colors and of advancing the colors to the chancel and retiring them is most impressive. The Jewish cadets had their service in the auditorium at 9:00 A.M. Sundays. Other denominations have periodic communion after the 9:00 A.M. service. For several years, the YMCA had religious instruction on Monday nights, but the chaplains of the different denominations preferred to have denominational groups meet under their instruction, except once a month, when combined meetings took place. Many weddings and a few baptisms have taken place in the chapel. For years, nearly all Protestant-Catholic weddings in Charleston were held in the chapel. Then a dispensation was granted for mixed marriages to be performed in the Catholic cathedral. Some, however, prefer the chapel.

There were no publicity, alumni, or placement activities when I came. But these I organized, and they became highly efficient. The number of secretaries increased with the need, and a corps of highly efficient women has existed for some years. The new library building and the purchase of stacks provided the physical requirements for the library. Money was liberally spent for books. A professional librarian and able assistants were employed, and the library met all needs. The laundry was largely expanded in buildings and machinery, and it has been operated by highly efficient personnel. The new laundry was a great improvement. The mess was noted for an abundance of good food. With a capable steward and a skilled dietician, it was praised by cadets and visitors. The hospital has been ably conducted. From only one nurse, it increased to two nurses and a complete operating staff. The maintenance departments were directed by superior technicians, and the entire plant was served with efficiency and economy. A high state of morale prevailed.

When World War II came, the entire senior class was called to duty as officers. Many members of the other classes enlisted. The Corps of Cadets fell to an enrolment of about 125. The cadets were replaced by Army Specialized Training Program[12] soldiers, who, at one time, numbered nearly two thousand. The college also processed and tested some eighteen hundred recruits of different branches. The government paid for all services rendered. For several months after the war, the navy leased three barracks to house men who were awaiting their discharge. The college then accepted veteran students, who, at one time, numbered nearly fifteen hundred. About half of them lived in barracks. At first, those who lived in barracks were a great deal of trouble. After we issued regulations for them and veteran students were employed as proctors to enforce these regulations, the deportment of veteran students became excellent.

After World War II, the officers the army initially detailed for duty in the military department were unfit for such duty. When I wrote the professor of military science and tactics, who was also employed as commandant of cadets, a statement of his failure to perform a very simple duty, he quit without resigning as commandant, and all of the others immediately resigned as assistant commandants. This was tantamount to mutiny. Fortunately, I had all of them relieved at once and an able officer came as professor of military science and tactics and commandant of cadets. I never called on the new detail as his assistants to perform any duty in the commandant's department. Able retired officers and the commandant maintained excellent discipline and high morale.

Various echoes from my past have been heard during my years at the Citadel. A professor became desperately ill with pneumonia. The doctor told me that nothing more could be done for him and that his recovery was very doubtful. Later, he told me that penicillin might help, but there was none in this area, and he did not know how to obtain any. Penicillin was just being tried and could not be procured for use. He asked if I knew anyone in Washington who could help us secure a little. I knew of no one. I returned to my house from the office at about 5:30 P.M., after a day of anxiety and thought on the subject. Almost immediately after entering, the doorbell rang, and I found a nice-looking gentleman at the door. He said: "General, you do not know me, but I know you. I was a second lieutenant in the artillery of the Thirty-second Division. When we were relieved from the line in the Meuse-Argonne battle, we were in bad condition and much depressed. You were the corps commander, and you came and talked to us. You encouraged and heartened us and restored our morale. I have always wanted to thank you for what you did. I was flying through Charleston, and the plane is waiting an hour at the airport. I could not resist the opportunity to come and see you." I asked him what he was doing. He said that he was a research doctor in Washington. I asked him if he knew anything about penicillin. He said that he knew all about it. When I told him about the sick professor, he told me to call a number in Washington and ask for a certain doctor. The secretary said that the doctor was in conference and could not be disturbed. I asked her to give him my name and say that I must speak to him at once. He answered very cordially, and I told him about the sick professor and asked if we could get some penicillin. He said that the only person who could issue penicillin was a doctor in Boston who came to Washington once a week and that he would arrive at seven o'clock the next morning. He said that he would explain the case to him and that, if he thought penicillin would help, he would send some by plane. On arrival, the Boston doctor called the Charleston doctor and sent some penicillin by plane. The next morning, I found the professor eating breakfast and feeling strong. He said that, from the first injection, his condition changed and he felt entirely different. He soon recovered and remained in good health. Like bread cast on the waters, I have found good reactions from gestures that may have appeared small coming to me after many years. I have also found much ingratitude for important services to people.

On coming to Charleston, my dear wife and I identified ourselves with the Church of the Holy Communion,[13] where I was baptized and confirmed when a student at the Holy Communion Church Institute, now the Porter Military Academy. The church was in bad repair, and we contributed liberally and were active in having it reconditioned. After about a year, I was elected to the vestry and became the senior warden. But, in 1942, I felt compelled to resign because of differences over the expenditures of church funds.

Within a few months after our arrival, I was asked to become chairman of the Red Cross chapter. I found it in very bad condition. The meager salaries to staff members had not been paid for months, and there was no money in the treasury. It could not function. I advanced money for letterheads, stamps, etc., and some money was raised. Conditions improved, and it became a large business during World War II. Unfortunately, the national Red Cross interfered, and the regional office under whom we were placed consisted of very inferior personnel. Because it was no longer an organization of the community, I resigned in 1946. It had taken much of my time and money. I also became identified with the Community Chest and was chairman of some of its fund-raising campaigns. When it ceased to be supported, it was abolished. It was revived under the War Fund, but I declined to be a part of it. I was a member of the War Fund and the State Defense Council during World War II.

When I came to Charleston, the chamber of commerce was having difficulty in raising money. I was made chairman of its membership committee for the campaign, which was successful. I was then elected to the board of directors for one term. As chairman of the committee of the army and navy, I was instrumental in saving the Charleston Navy Yard, which had been ordered decommissioned. For a number of years, I was chairman of the committee on schools and colleges of the Tuberculosis Association, and, one year, I was state chairman for that campaign. I was also chairman of the campaign to raise money for the expansion of the YMCA. After World War II, I was chairman of the campaign to raise money for the Post-War Development Board and was a member of the board until I resigned in 1948. During the war, I was a member of the USO board and was active in the Seaman's Club in the Villa Marguerita. During most of the time, I was on the Salvation Army board and was chairman for a number of years before I left. I made many addresses, attended many banquets, introduced or thanked speakers, installed officers of the chamber of commerce and the regional Boy Scouts, and met dignitaries.

The most important event in my life in Charleston was when I was made a Mason at Sight in 1935. I at once affiliated with Pythagorean Lodge and became, in turn, senior deacon, senior warden, and master. I was then chaplain for several years. As soon as I became a Mason, I joined the Scottish Rite and became an officer in some of its bodies. I was also made a member of the executive committee, which managed the business. In 1936, I was appointed deputy of the Supreme Council in South Carolina and, in 1937, received the thirty-third degree and was elected to the Supreme Council and made sovereign grand inspector general in South Carolina. The meetings of the Supreme Council in Washington were interesting events for my dear wife and me. She took her full share in entertaining the ladies at the Army and Navy Club, and all became devoted to her. After her first stroke in 1942, she could do very little.

I was a delegate from the parish to the diocesan conventions and, from the diocese, to the national conventions of the Episcopal Church. For some years, I was a member of the State Rural Electrification Authority and was active in extending rural electrification until it was taken over by county cooperatives.

In 1932, I was asked to become chairman of the Florida Ship Canal Authority to construct a ship canal across Florida. I saw its advantages for the national defense and accepted. I was also chairman of the Ship Canal Navigation District, which procured land and received taxes for the canal. We secured about $7 million from the federal government, and much work was done. But the railroads fought this with a powerful lobby, in and out of Congress, and managed to have work suspended for lack of funds. We were successful in having money authorized but could not get it appropriated by Congress. During World War II, it would have saved many times its cost by allowing ships to pass between the Atlantic and the Gulf without being sunk by submarines. It would have transported all the oil required, and in many ways it would have been invaluable. In 1953, the authority ended my chairmanship with no acknowledgment of my services to the state.

As president of the Society of the First Division, I attended the annual reunions all over the country and did much to preserve the unity and high morale of the men. I also attended the officers' annual dinner in New York, with good results to all. In 1953, they rendered me a special honor. I also was president of the First Division Memorial Association, which raised funds for erecting the monument in Washington, and did much to preserve the other monuments in Washington and in France and to safeguard our endowment fund for their maintenance in perpetuity. For all these activities, I received a number of trophies and testimonials, which I value highly.

At the beginning of World War II, I offered my services to the War Department, realizing that I would not be desired. I had a curt acknowledgment, which was all that I expected. With my experience and knowledge of the services, I could have been of great value, but it would have been at a large financial sacrifice. I know that I was rendering valuable service by training efficient officers for the army. I know that my war plans would have prevented Pearl Harbor unless the fleet was deliberately placed there for the purpose of having it destroyed in order to force the country into the war. When I learned that the fleet was in Pearl Harbor, I knew that it was doomed. Senator Maybank[14] quoted me as saying so long before the disaster. When my daughter-in-law saw me walking the floor one night, she asked what was the matter. I told her that I had just learned that the fleet was in Pearl Harbor and that it was doomed.

The Citadel continued to absorb my interest, and I was anxious to complete my construction program, for which funds were available. After some correspondence and discussion with the chairman of the Board of Visitors as to my resigning as president of the Citadel, we agreed that I would give the Board of Visitors ample notice when I decided to leave. The chairman of the board[15] then wrote me that, in order to secure my successor, it was necessary for me to fix a date for me to leave. I gave the date of June 30, 1953, for my resignation to take effect. On that day, I left the Citadel and took up my residence at Whitehall, Aiken, South Carolina. The Citadel faculty and staff gave me a banquet and other testimonials of their loyalty and esteem.

Whatever success I had at the Citadel is due in the greatest measure to the loyal support and untiring labors of a number of officers, staff, and employees. Colonel J. W. Lang, by his ability, force, and devotion to duty, was responsible for the high standards of duty and discipline that eventually prevailed. Colonel McMurray did much to eliminate evils by his high courage and conscientious performance of duty. Colonel C. F. Myers[16] was an outstanding leader in administration and procurement. All heads of departments and their assistants maintained high academic standards. Mr. F. P. Kanapaux[17] operated the engineering plants, water, and sewage with rare skill.

Chapter 30

The Men Who Influenced My Life

Of course, my father, my example and guide in childhood and boyhood, shaped my ideas and conduct in the largest measure. He was practically self-educated. I never heard him refer to going to school. He had a fine mind and was well informed on medicine, law, business, and many skills like wheelwright work, carpentering, building, painting, blacksmithing, and farming. I never knew a man with such varied knowledge. He wrote a good hand and used good English. He read many worthwhile books and had sound ideas on politics, religion, and government. He inspired confidence and made friends with the best people, but he incurred the bitter enmity of an outlaw class, who could not influence him. He had a great moral courage and expressed himself freely on all important matters. He was justice of the peace and postmaster in Providence. His especial bent was medicine, and he was frustrated because he could not be a doctor. He gave his family medical care in all except serious cases. Yet he had the greatest difficulty in making a living, and, but for the money earned by my mother in all of my childhood and boyhood in teaching short terms of schools for very small pay, I do not see how we would have lived.

My mother's influence was even greater. She made all our clothes and, with a little help from a colored woman, did all the housework. After I was eleven years old, she did all the work with my help. I did much in cooking and washing and cut wood, built fires, etc. She taught us all we knew in books, about religion and the Bible, and influenced us to read good books. My character and my habits of thought, action, and speech were entirely formed by my parents. My mother's intellect, education, and breeding gave us high standards of behavior and a consciousness of noblesse oblige. We knew that we were descended from honorable and well-bred people who earned distinction in various ways by their patriotism, their civic achievements, and their military service. Our poverty never lowered our pride, and we were respected and liked by the best people.

To the Reverend A. Toomer Porter I owe the opportunity to rise above what seemed to be hopeless poverty and humble living. On my mother's plea, he took me to his school, the Holy Communion Church Institute in Charleston, South Carolina, and gave me an education without charge. His fees for books, board, and tuition were $210 per year, but very few of the boys could pay anything. We never knew about one another's financial status, and we were at no disadvantage because of poverty. The railroads and steamships gave us passes through Dr. Porter, and our parents managed to pay the uniform fee of $12 and for one suit of clothes, which I wore every day. During my third year, my brother was teaching in the school, and he paid my fee and bought me my first overcoat. In after years, I was able to pay Dr. Porter in full and to return to my brother his assistance many times over. At Dr. Porter's, I was baptized and confirmed in the Episcopal Church, which has had a great influence on my life. Religious opportunities, except occasional Sunday schools, were few in my boyhood.

While I never saw him but the one time, I really owe my appointment to West Point to Mr. Sherman Conant, a leading railroad builder in Florida. I had won the competitive examination in Jacksonville, thanks to my education at Dr. Porter's, but Congressman Charles Dougherty would not give me the appointment. Mr. Conant had sent him to Congress, and Dougherty had made him a member of the straw board for the competition. Just before it was time for me to leave for West Point for the entrance examination, I met Mr. Conant in Leesburg and told him my situation. He was very angry and said: "That man cannot treat you this way." In a few days, I received the appointment.

Shortly after I joined the First Infantry at Benicia Barracks, California, I met the daughters of William Montrose Graham, who, as colonel, commanded the Fifth Artillery at the Presidio of San Francisco. I met him when I called on the young ladies, and he was always very courteous to me. When I decided to transfer to the artillery, I went to him, and he really influenced my transfer with the authorities in Washington. He was an uncompromising disciplinarian and demanded the highest efficiency. He was greatly feared by officers and soldiers, but he always treated me kindly. By appointing me as quartermaster and commissary for the troops in the field during the railroad strikes in 1894 and, later, appointing me regimental adjutant, which the War Department made him revoke on account of my junior rank, he gave me a prestige that had far-reaching results. When he was made a major general in 1898, he made me his aide, where I had the opportunity to attract attention to my work. When General Graham retired in 1899, I was asked by General A. C. M. Pennington to become his aide in Atlanta, Georgia. I owe General Graham much for his favor and esteem.

In May 1899, Reilly's Battery was ordered to the Philippines, and he had me assigned to the battery. This gave me the opportunity for service in the Philippines and China, where I got my head out of the crowd. From then on, I had an individual reputation and standing in the army. This led to my details at Fort Myer and West Point, where my work became widely known. Captain Reilly had a great influence on my career.

In the Philippines, I came in contact with General Robert L. Bullard, who took a fresh regiment of volunteers to Calamba, where I was serving with the Twenty-first Infantry. He was a colonel and had never heard of me. His first act was to relieve me of some guns manned by infantry, but Colonel Kline of the Twenty-first Infantry had him revoke the order. We immediately had some fighting with the insurgents, and Colonel Bullard became very friendly. Before the campaign was ended, I was bound to him for life. He kept occasional contact with me until the world war. When he was assigned to command the First Division, he had me transferred from the Forty-second Division, where I commanded the Sixty-seventh Artillery Brigade, to the First Field Artillery Brigade of the First Division. This gave me my start in the AEF. His confidence and loyalty to me never wavered, and, when he was promoted to the command of a corps, he, no doubt, did what he could to have me promoted and assigned to the command of the First Division. I owe to him more than anyone else my advancement in the AEF.

To President Calvin Coolidge I am indebted for becoming chief of staff. I had met him when he was the governor of Massachusetts. He sent me a copy of his book *Have Faith in Massachusetts*. I telegraphed him my congratulations when he was elected vice president and when he became president. I had every reason to believe that the officers associated with the AEF GHQ did everything possible to have one of its staff appointed. As far as I know, I had no loyalty or support from any of this group while I was in office. President Coolidge became a strong friend even after two very serious differences over the budget. He was brave and honest, and, when he saw my side, he gave me his full support. I owe him my promotion to full general against the hardest fight that the navy could make.

No one has been more helpful in sustaining me in difficulties and in accomplishing worthwhile purposes than the late Colonel Clark Williams. I met him when he was the director of the Red Cross in the First Division and at once recognized his superior heart, mind, and spirit. In all the years that followed, his helpfulness, his wise counsel, and his unswerving loyalty have meant more than he could know or I could say. His twenty-one scholarships at the Citadel are only one evidence of his great heart.

Among my staunchest friends and supporters is Colonel Robert R. McCormick. From the time that I met him as a major commanding a battalion of the Fifth Field Artillery, I have relied on him in every way, and much that I have accomplished is due to his loyalty, ability, influence, and resources. His courage, devotion to duty, technical skill, and leadership were outstanding in the operations in Lorraine and Picardy. When I needed publicity for the Citadel, he gave space in the *Chicago Tribune* that greatly increased enrollment. For a number of years, he has given two scholarships annually to the Citadel, one recipient later sacrificing his life in World War II. The home to which I retired on leaving the Citadel was a munificent gift to the college by him for my use.

General Hugh L. Scott, the superintendent of the military academy, became a real and loyal friend at West Point. It was due to him that my detail as senior instructor of artillery tactics was extended twice, with the result that I was on duty there for five and a half years. He supported me in every way and was most helpful in obtaining guns, base-end stations, searchlights, and complete fire control equipment for all artillery instruction. His friendship continued until his death.

To the chiefs of staff of all my active commands I owe a great measure of my success. General Campbell King of the First Division, General W. B. Burtt of the Fifth Corps, and General Briant Wells of the Fourth Corps and, later, deputy chief of staff in the War Department were able, loyal, and highly trained officers, and they gave me the full measure of their talents. I have deeply regretted that I could not reward them all as I so much wanted to do. My gratitude to them is unbounded.

To the officers and soldiers of the First Division I owe my greatest measure of success. Their loyalty and devotion during World War I and in all the years since are more than any man could deserve. I believe that no one has ever been so much honored as I have by their admiration and confidence.

I owe boundless gratitude to great numbers of officers and soldiers who gave me the most able and loyal support in carrying out my policies and in advising and informing me. I could have done nothing without them. If I began to list these names, I would leave out many, however long the list might me. I tried to tell them at the time by letters, citations, and verbal expressions of my gratitude.

As though the Lord had sent him, Grand Commander John H. Cowles, 33° of the Scottish Rite, opened a new world to me in Masonry. His friendship and confidence more than anything else advanced me to be a member of the Supreme Council, which is one of the greatest honors that a man can have.

While the list of my enemies is long, it is futile to name them. My conscience is clear in my dealings with them. I never wronged one knowingly, and I was instrumental in advancing some of them. The two outstanding causes were jealousy of my success and resentment because I would not allow them to continue to violate the law by being reappointed chiefs of branches, thus depriving other officers of their right to the offices.

Notes

Editor's Note

1. At about the time Summerall retired from the army, the Houghton Mifflin Co. contacted him, encouraging him to "write [his] Reminiscences," which would "find a waiting public." See Roger L. Scaife to Charles P. Summerall, December 10, 1930, box 8, folder September 1930–August 1931, S–Z, Charles P. Summerall Papers, Manuscript Division, Library of Congress (hereafter Summerall Papers). But he never wrote for publication.

Introduction

1. Biographical information on Summerall and his military service can be found in Larry Addington, "Charles Pelot Summerall," in *Dictionary of American Military Biography,* ed. Roger J. Spiller (Westport, CT: Greenwood, 1984), 1077–80; Charles L. Anger, "Charles Pelot Summerall," in *Dictionary of American Biography: Supplement 5* (New York: Scribner's, 1977), 668–69; Charles P. Summerall Jr., "Charles P. Summerall," *Assembly,* January 1956, 57–58; and Timothy K. Nenninger, "Charles P. Summerall as Chief of Staff of the U.S. Army, 1926–1930" (1986, typescript). Summerall's pre–World War I consolidated personnel file is 3672-ACP-1892, Appointment, Commission, and Personal Branch, Adjutant General's Office, Record Group (RG) 94, National Archives (NA).

2. Colonel Robert L. Bullard to the Adjutant General, March 15, 1900, 3672-ACP-1892, RG 94, NA.

3. Efficiency Report by Colonel A. P. Hatfield, June 30, 1905, and Efficiency Report by Colonel H. L. Scott, June 30, 1910, 3672-ACP-1892, RG 94, NA.

4. Report of the Artillery Section of the American Military Mission, June–July 1917, box 26, Summerall Papers.

5. Addington, "Charles Pelot Summerall," 1079; Fletcher Pratt, *Eleven Generals: Studies in American Command* (New York: William Sloan, 1949), 250–54.

6. Donald Smythe, "A.E.F. Snafu at Sedan," *Prologue,* Fall 1973, 135–49.

7. For their mutual admiration society, see John J. Pershing to Charles P. Summerall, October 28, 1918, and Summerall to Pershing, October 31, 1918, box 193, John J. Pershing Papers, Manuscript Division, Library of Congress.

8. Liggett's changing attitude is reported in the October 11, 1918, November 8, 1918, and May 6, 1919, diary entries of his aide, Lieutenant Colonel P. L. Stackpole. The diary is held at the George C. Marshall Library, Lexington, VA.

9. Duffy quoted in D. Clayton James, *The Years of MacArthur,* vol. 1, *1880–1941* (Boston: Houghton Mifflin, 1970), 222.

10. Russell F. Weigley, *History of the United States Army* (New York: Macmillan, 1967), 403.

11. Information from this section comes from Nenninger, "Charles P. Summerall as Chief of Staff."

12. For more on Summerall, especially as Citadel president, see W. Gary Nichols, "The General as President: Charles P. Summerall and Mark W. Clark as Presidents of the Citadel," *South Carolina Historical Magazine* 95 (October 1994): 314–35, and "General Charles P. Summerall: The Training, Command, and Education of the Citizen-Soldier," in *Unknown Soldiers: The American Expeditionary Forces in Memory and Remembrance,* ed. Mark A. Snell (Kent, OH: Kent State University Press, 2008), 144–64.

13. B. C. Mossman and M. W. Stark, "Former Chief of Staff General Charles P. Summerall, Combined Services Full Honors Funeral, 14–17 May 1955," in *The Last Salute: Civil and Military Funerals, 1921–1969* (Washington, DC: Department of the Army, 1991), 87–88.

1. The Rock Whence I Was Hewn

1. Private, Company E, Ninth Florida Infantry.

2. In February 1864, Brigadier General Truman Seymour, commanding mostly U.S. Volunteer troops from Massachusetts, occupied Jacksonville and cleared northern Florida of Confederate forces.

3. In 1598, the Treaty of Nantes granted French Huguenots equal political rights with Catholics but did not secure them complete freedom of worship.

4. In 1731, Purysburg, on the Savannah River lowlands, was settled by six hundred poor Swiss immigrants who wanted to establish a silk trade in the area, but the effects of malaria and other hardships proved impossible for them to overcome. The settlement died a slow death and was abandoned in the 1830s.

5. The capture of the *Waterwitch* is documented in U.S. War Department, *The War of the Rebellion: A Compilation of the Official Records of the Union and Confederate Armies and Navies,* 4 ser. in 128 vols. (Washington, DC: U.S. Government Printing Office, 1880–1901), ser. 1, 15:475–510.

6. U.S. Naval Academy (USNA) 1892; resigned 1907 as a lieutenant commander; died 1935.

2. The Pit Whence I Was Digged

1. A type of palm, with four shaped leaves.

2. Porter lived from 1829 to 1902. The school he founded in 1867 became know as the Porter Military Academy late in the nineteenth century and continues today as the Porter-Gaud School in Charleston. His memoir was *Lead On! Step by Step: Scenes from Military, Educational and Plantation Life in the South* (New York: Putnam's, 1898).

3. Cleveland was the first Democratic president since the Civil War and the only one between 1860 and 1912.

4. Charles Dougherty of Port Orange was the Second District congressman. A Democrat first elected in 1885, he served until 1889.

3. "And David was wise in all his ways and the Lord was with David"

1. William Cary Brown; U.S. Military Academy (USMA) 1877; retired 1919 as a brigadier general; died 1932.

2. William F. Spurgin; ex-USMA 1862; retired 1902 as a brigadier general; died 1904. The use of *ex-* in this context indicates that Spurgin entered West Point with the class of 1862 but did not graduate. Many such failed cadets received army commissions through other means.

3. During their first summer, first-year cadets, plebes, are indoctrinated to the rigors of academy life.

4. Isaiah 51:1.

5. Died 1895 as a second lieutenant.

6. One of Summerall's best friends in the army; retired 1926 as a brigadier general; returned to active duty 1941–46 as an adviser to the chief of staff on manpower policy; died 1955. See I. B. Holley Jr., *General John M. Palmer: Citizen Soldiers and the Army of a Democracy* (Westport, CT: Greenwood, 1982).

7. "Fresh, lacking in respect. Bold before June" (1942 *Bugle Notes*). "Before June" means before a plebe becomes an upperclassman.

8. Equivalent of a college sophomore.

9. Killed in action in 1898 at San Juan Hill while serving as a first lieutenant with the Seventeenth Infantry.

10. The experience of the class of 1892 is nicely told in George S. Pappas, "Turn Back the Clock to 1892," *Register of Graduates and Former Cadets, 1992* (West Point, NY: Association of Graduates, 1992), 11–23, 33–41, 925–31.

11. Unable to identify the officer to whom Summerall refers.

12. The West Point "color line" was the occasion when plebes were induced by upperclassmen to perform skits and provide other entertainment, usually at summer camp.

13. Charles Young; USMA 1889; died 1922 as a colonel; was the third black graduate of West Point. The life of black cadets at West Point during this period was not easy. Between 1870 and 1889, there were twelve black cadets, but only six stayed longer than a semester. Young, who had difficulty with mathematics, took five years to graduate, but his perseverance won admiration from his classmates and other cadets. See Edward M. Coffman, *The Old Army: A Portrait of the American Army in Peacetime, 1789–1898* (New York: Oxford University Press, 1986), 226–29.

14. USMA ex-1891.

15. Stood at an exaggerated position of attention.

16. Ranald MacKenzie (1840–1889); USMA 1862; retired 1884 as a brigadier general; was an acclaimed cavalry leader and Indian War campaigner.

17. "Richly dressed with tack."

18. A poem written by Cecil Frances Humphreys Alexander (1818–1895), an Irish poet who often wrote on religious themes.

19. The regular army officers responsible for the discipline and military training of cadets.

20. The ranking cadet in the Corps of Cadets.

21. Dance.

22. McCormick (1880–1955), the longtime publisher of the *Chicago Tribune,* served as an artillery commander with the First Division in World War I, where he became a friend and colleague of Summerall's. See Richard Norton Smith, *The Colonel: The Life and Legend of Robert R. McCormick* (Boston: Houghton Mifflin, 1997).

23. The assignment of cadets to ranks and billets in the Corps of Cadets military chain of command.

24. USMA 1892; retired 1921 as a colonel; died 1950.

25. Henry Howard Whitney; USMA 1892; retired 1930 as a brigadier general; died 1949.

26. Born 1858; was a playwright and comedian in New York City.

27. USMA 1892; died 1930 on active duty as brigadier general.

28. The first army-navy football game occurred in the fall of 1890. The more experienced navy team won 24–0, although army won the next year 32–16.

4. "We bid farewell to cadet gray and don the army blue"

The chapter title references the opening lines of the song "Army Blue," the words (by George T. Olmstead, L. W. Becklaw, et al.) sung to the tune of George R. Poulton's "Aura Lee."

1. Captain Leopold O. Parker; second lieutenant U.S. Volunteers 1865; retired 1904 as a lieutenant colonel.

2. William M. Wright; USMA ex-1886; retired 1922 as a lieutenant general; died 1943. He commanded the Eighty-ninth Division in World War I, for a time as part of Summerall's Fifth Corps. See William M. Wright, *Meuse-Argonne Diary: A Division Commander in World War I*, ed. Robert H. Ferrell (Columbia: University of Missouri Press, 2004). Wright held command of a number of corps during and immediately after the war, but not in combat—First Corps, November 13, 1918–February 28, 1919; Third Corps, June 17–July 12, 1918; Fifth Corps, July 12–August 19, 1918; and Seventh Corps, August 19–September 6, 1918.

3. A four-wheeled carriage without a top.

4. USMA 1877; retired 1919 as a major general; died 1919.

5. Leopold O. Parker.

6. In 1891, John M. Schofield, the commanding general of the army, required officers of all line garrisons to meet regularly to discuss professional subjects and prepare papers on military topics.

7. In December 1890, the battle at Wounded Knee was the last major encounter between the army and the Indians.

8. Second Lieutenant, Artillery, 1855; retired 1898 as a brigadier general.

9. From the Civil War until World War I, the Medal of Honor was the only decoration for valor. Brevet ranks, above the regular commission, were awarded to officers for meritorious or gallant service and resulted in additional pay and recognition. For a discussion of brevets, see Robert M. Utley, *Frontiersmen in Blue: The U.S. Army and the Indian, 1848–1865* (New York: Macmillan, 1967), 32–33.

10. In the immediate post–World War I period, the Class B system employed evaluation boards to eliminate overaged and inefficient officers.

11. Captain Henry J. Reilly; enlisted service 1864–1866; second lieutenant, Fifth Artillery 1866; killed in action in China 1900 as a captain; was one of Summerall's mentors and lifelong role models.

12. A timepiece capable of accurately measuring extremely brief intervals of time.

13. An increase in base salary based on length of service.

14. Horse-drawn artillery, armed with three-inch guns that fired fifteen-pound projectiles.

15. "Mileage money" is essentially expense account funds provided to soldiers to cover their travel costs.

16. The recently formed American Railway Union conducted a short and successful strike against the Great Northern Railroad in April 1894. Then almost immediately began a strike against the Pullman Co. that affected nationwide railroad operations. For more on the army and the 1894 labor strikes, see Jerry M. Cooper, *The Army and Civil Disorder:*

Federal Military Intervention in Labor Disputes, 1877–1900 (Westport, CT: Greenwood, 1980), 99–143.

17. USMA 1887; died 1899 at Manila as a lieutenant colonel.

18. Laura Modecai Summerall (June 24, 1872–April 23, 1948); married Summerall on August 14, 1902.

19. Alfred Mordecai, USMA June 1861; retired 1904 as a brigadier general; died 1920.

20. USMA 1892; retired 1932 as a brigadier general; died 1936.

21. Second Lieutenant George G. Gatley; USMA 1890; died 1930 as a colonel; promoted posthumously to brigadier general.

5. Remember the *Maine*

1. The ship had been sent to Cuba to show the flag in support of Cubans seeking independence from Spain. The ship's destruction, resulting in the death of 253 sailors, was the proximate cause of the United States going to war against Spain, with public opinion blaming the Spanish for attacking the ship. Subsequent investigations reached a more ambiguous conclusion. An exhaustive 1976 study, supervised by Admiral Hyman Rickover, concluded that an accidental internal explosion, not a Spanish mine or torpedo, had sunk the *Maine*. See Kenneth J. Hagan, *The People's Navy: The Making of American Sea Power* (New York: Free Press, 1991), 212–14.

2. Lieutenant Colonel Joseph G. Ramsay; second lieutenant, Second Artillery 1861; died 1899 as a lieutenant colonel.

3. At Dunn Loring in northern Virginia, near Washington, DC.

4. Matthew Calbraith Butler (1836–1909); rose from captain to major general in the Confederate army; served as a Democratic senator from South Carolina 1877–1895; was a major general of the volunteers during the Spanish-American War.

5. John P. S. Gobin (1837–1912); Union army brevet brigadier general; lawyer, political leader, and lieutenant governor of Pennsylvania.

6. Colonel William Nalle commanded the Third Virginia Infantry in 1898. But neither his compiled military service record (RG 94, NA) nor the index to general courts-martial (RG 153, NA) provide information about a court-martial.

7. William R. Shafter; first lieutenant Michigan Volunteers 1861; retired 1899 as a brigadier general; died 1906; promoted posthumously to major general. In 1898, Shafter was promoted to major general of the volunteers and given command of the force sent to Cuba. Although the command relatively easily defeated the Spanish troops on the island, it suffered serious logistic and medical problems, for which Shafter received much blame.

8. Alexander C. M. Pennington; USMA 1860; retired 1899 as a brigadier general; died 1916.

6. The Little Brown Brother

Although offensive to twenty-first-century sensibilities, the chapter title is Summerall's, the product of another time, place, and frame of mind. Used on occasion to demean, as often as not the expression reflected the peculiar paternalism that Americans in the islands felt for Filipinos.

1. The best books on the U.S. Army in the Philippines are Brian M. Linn, *The U.S. Army and Counterinsurgency in the Philippine War, 1899–1902* (Chapel Hill: University of North Carolina Press, 1989), and *The Philippine War, 1899–1902* (Lawrence: University Press of Kansas, 2000).

2. Louis Ray Burgess; USMA 1892; retired 1934 as a colonel; died 1938.

3. USMA 1898; retired 1938 as a brigadier general; died 1963.

4. Captain Sydney W. Taylor, Battery F, Fourth Artillery; commissioned regular army 1867; retired 1911 as a colonel; died 1941.

5. John L. Tiernon, First Artillery; commissioned 1862 from Missouri; retired 1903 as a brigadier general; died 1910.

6. Summerall is referring here to the controversial tactics, high casualties, and slow progress that characterized the British effort in the Boer War, 1899–1902.

7. Major General Elwell S. Otis; Harvard Law School graduate; Civil War volunteer service; commanding all U.S. troops in the Philippines at this time; retired 1903 as a brigadier general; died 1909.

8. USMA 1890; retired 1931 as a major general; died 1941.

9. According to the 1899 and 1900 *Army Registers,* there were seven first lieutenants in the Twelfth Infantry during this period who were at West Point during the time Summerall was a cadet, although none was in his class. I am unable to determine to which individual he refers.

10. Colonel Jacob Kline; commissioned 1861 as a second lieutenant in the Pennsylvania Volunteers; retired 1904 as a brigadier general; died 1908.

11. USMA 1894; retired 1935 as a brigadier general; died 1937.

12. This was a kind of machine gun, based on a Gatling gun design.

13. USMA 1876; retired 1918 as a brigadier general; died 1934.

14. Second Lieutenant Paul Stockley; commissioned regular army 1899; killed in action 1899 in the Philippines.

15. Explosive gelatin or dynamite was used to increase the range and power of the projectiles.

16. Robert Lee Bullard; USMA 1885; retired 1925 as a major general; promoted 1930 to lieutenant general on the retired list; was a brigade,

division, corps, and army commander in World War I; died 1947. See Allan R. Millett, *The General: Robert L. Bullard and Officership in the U.S. Army, 1881–1925* (Westport, CT: Greenwood, 1975). Although a few years older, Bullard was probably Summerall's closest professional colleague.

17. Miguel Malvar (1865–1911) was the *insurrecto* leader in Batangas Province.

18. B. Frank Cheatham; was a Tennessee Volunteer officer; commissioned regular army 1900; later served in the Quartermaster Department; retired 1930; died 1944.

19. George Smith Anderson; USMA 1871; retired 1911 as a brigadier general; died 1915.

20. *Insurrectos* armed with large, single-edged knives or machetes.

21. On the organization and use of the Macabebes, a Filipino tribe friendly to the United States, see Linn, *The Philippine War* (2000), 128.

22. For more on Hall's (1849–1911) interesting career, see Dora N. Raymond, *Captain Lee Hall of Texas* (Norman: University of Oklahoma Press, 1940).

23. A rope with hooks to pull gun carriages.

7. The Land of the Dragon

1. A recent, well-researched account is Diana Preston, *The Boxer Rebellion: The Dramatic Story of China's War on Foreigners That Shook the World in the Summer of 1900* (New York: Walker, 1999).

2. Two informative manuscript sources on the Fifth Artillery in the Philippines and China are First Lieutenant Louis R. Burgess, Report on the Operations of the Light Battery, Fifth Artillery, in China, August 17, 1900, and Historical Sketch of the 10" Battery, Field Artillery, on Foreign Service, 1899–1900, both in box 2, Summerall Papers.

3. The practice of using "prisoners," soldiers under punishment for disciplinary infractions, for "fatigue duty" and other work details was widespread in the army for much of the twentieth century.

4. Emerson H. Liscum; commissioned 1863 from Vermont; killed in action 1900 at Tientsin as a colonel.

5. Adna Romanza Chaffee; commissioned regular army 1863, Sixth Cavalry; was the commander of the U.S. contingent in the China Relief Expedition; retired 1906 as a lieutenant general; died 1914.

6. Lieutenant General N. P. Linievitch, the commander of Russian troops in the expedition; died 1908.

7. Parapets with regular, repeating, squared indentations.

8. George P. Scriven; USMA 1878; retired 1917 as a a brigadier general; died 1940.

9. Littleton Waller Tazewell Waller commanded a battalion of marines in the expedition; commissioned 1880; retired 1920 as a major general; died 1926.

10. William Crozier; USMA 1876; retired 1919 as a major general; died 1942.

11. Troops from Kansu commanded by the Muslim general Tung Fu-hsiang constituted one of the elements of the Boxer insurgents.

12. For more on the depredations against innocents, see Preston, *The Boxer Rebellion*, 276, 303, 304n.

13. Enamel artwork with thin metal bands separating patterns.

14. The Chinese high commissioner and minister plenipotentiary, friendly to Western interests.

15. USMA 1883; retired 1919 as a colonel; died 1939; the father of General Matthew B. Ridgway.

16. B.S. Princeton University; commissioned 1898 from Ohio; died 1932 as a colonel.

17. On July 1, 1901, the army promoted Summerall to captain.

18. Inchon.

19. Field Marshal Count Alfred von Waldersee (1832–1904), commanding Allied forces in China.

8. Back to Manila and Home

1. General Alfred Gaselee (1844–1918) commanded the British contingent in China.

2. Lieutenant Colonel Theodore J. Wint; commissioned Pennsylvania Volunteers 1864; died 1907 as a brigadier general.

3. Captain Bernard Kelly; was a Civil War volunteer officer; commissioned regular army 1897 as a chaplain; retired 1902; died 1926.

9. The Land of the Midnight Sun

1. Many army doctors in the late nineteenth century and the early twentieth were not regularly commissioned officers but had a term contract defining the duties they would perform.

2. Ever since the 1825 treaty between England and Russia concerning boundaries in what became Alaska's southern panhandle, it was unclear which territory in the area belonged to the United States and which to England. With the United States having unfettered access to routes leading to the Klondike goldfields of importance to Canada, the dispute over boundaries became a point of contention between the two countries. In 1903, an international commission, meeting in London, decided the dispute in favor of the U.S. claim. See "Alaska Boundary

Question," in *Concise Dictionary of American History,* ed. Thomas C. Cochran (New York: Scribner's, 1962), 27–28.

3. 1902–1979; USMA 1924; commanded the Ninety-first Field Artillery Battalion, First Armored Division, in World War II; retired 1954 as a colonel.

4. William A. Wickline; appointed an assistant surgeon 1904; retired 1933 as a colonel; died 1974.

5. Powder so fine that, when rubbed between the fingers, no grit is felt.

6. 1856–1944; adventurer, businessman, and freighting entrepreneur in Alaska.

7. Jefferson Randolph Smith II (1860–1898); confidence man, gambler, and saloon proprietor in Alaska during the gold rush days.

8. Brigadier General George M. Randall; commissioned 1861; retired 1905 as a major general; died 1918.

9. USMA 1884; retired 1920 as a colonel; died 1929.

10. The Coast Defenses

1. Erastus Corning Hawkins (1860–1912) was an internationally known construction engineer who moved to Alaska in 1898; he was vice president, chief engineer, and general manager of the White Pass and Yukon Railroad.

2. Colonel George S. Grimes; commissioned 1862; retired 1907 as a brigadier general; died 1920.

3. Frederick Funston; commissioned 1898 as a colonel in the Nebraska Volunteers; appointed regular army brigadier general 1901; died 1917 as major general.

4. George E. Pickett; USMA 1846; major general, CSA; died 1875.

5. A passing reference to the 1843–1846 boundary dispute characterized at the time as "fifty-four forty or fight." In the end, President James Polk compromised with England and established the border on the forty-ninth parallel.

6. John W. C. Abbott; commissioned 1899 from Nebraska; retired 1920 as a lieutenant colonel; died 1940.

11. The Caissons Go Rolling Along

1. Eli DuBose Hoyle; USMA 1879; retired 1915 as a brigadier general; died 1921.

2. Major General Joseph Wheeler (1836–1906) was the principal Confederate cavalry commander in the western theater during the Civil War; he also served as a major general of volunteers during the Spanish-American War and the Philippine Insurrection.

3. First Sergeant Alfred F. Hart, Stable Sergeant Charles C. Vite, and Sergeant Charles E. Kelly, Third Battery, Field Artillery.

4. Major General Alfred Pleasonton; USMA 1844; major general of volunteers and Civil War Union cavalry corps commander; retired 1888; died 1897.

5. Summerall is probably referring here to Colonel Frank Smith Robertson (1841–1926), who was an engineer officer on Jeb Stuart's staff during the Civil War. See Robert J. Trout, ed., *Into the Saddle with Stuart: The Story of Frank Smith Robertson of Jeb Stuart's Staff* (Gettysburg, PA: Thomas, 1998).

6. Lewis H. Strother; VMI 1877; was commandant of cadets at the institute 1903–1905; died 1908.

7. Scott Shipp; USMA 1859; was VMI superintendent 1889–1907; died 1917. About this time, Shipp tried to recruit Summerall for the VMI faculty, but the request was denied by the secretary of war. Captain W. M. Wright to Charles P. Summerall, June 18, 1906, box 2, Summerall Papers.

8. Samuel Houston Letcher (1849–1914), among the VMI cadets who fought against Union troops at New Market in 1864, was a Virginia circuit court judge.

9. Stephen M. Foote; USMA 1884; died 1919 as a brigadier general.

10. USNA 1863; retired 1908 as an admiral; died 1912; among other duties commanded the Asiatic Fleet and the "round the world" Great White Fleet.

11. Unable to identify this officer.

12. Edward P. Nones; USMA 1900; died (drowned) 1916 as a captain.

13. Henry S. Kilbourne Jr.; USMA 1903; retired 1930 as a colonel; died 1967.

12. To West Point

1. The military academy artillery detachment consisted of regular army soldiers who provided real-world instruction to cadets on the care, employment, and capability of artillery and its associated equipment and horses.

2. Lieutenant Colonel Robert L. Howze; USMA 1888; died 1926 as a major general.

3. Captain Merch B. Stewart; USMA 1896; retired 1925 as a brigadier general; died 1934.

4. Earl Horatio Herbert Kitchener (1850–1916); British army field marshal.

5. General Baron Tamemoto Kuroki (1844–1923), who commanded the First Japanese Army during the Russo-Japanese War, was on a tour of U.S. Army installations at this time.

6. George S. Patton Jr. (1885–1945); USMA 1909; commanded Third Army in Europe in World War II. Henry H. Arnold (1885–1950); USMA 1907; commanding general of the army air forces in World War II. Carl Spaatz (1891–1974); USMA 1914; commanding general of the U.S. Strategic Air Forces in Europe during World War II. Jacob L. Devers (1887–1979); USMA 1909; commanded the Sixth Army Group in Europe in World War II. Alexander H. Patch (1889–1945); USMA 1909; commanded the Seventh Army in Europe in World War II. William Hood Simpson (1888–1980); USMA 1909; commanded the Ninth Army in Europe during World War II. Robert L. Eichelberger (1886–1961); USMA 1909; commanded the Eighth Army in the Pacific during World War II. Simon Boliver Buckner (1886–1945); USMA 1908; commanded the Tenth Army in the Pacific during World War II; killed in action on Okinawa.

7. Robert L. Howze.

8. Henry MacKay was a manufacturer of rugs and linoleum. He lived in Brooklyn, not on Long Island. One daughter married William R. Henry, the other Edwin C. McNeil, both army officers, both USMA 1907. MacKay died in 1939. Summerall clearly was fond of the MacKays, and it is possible, but not verified, that this is the "wealthy family in New York" to which he refers in his early chapter on West Point.

9. USMA 1892; retired 1932 as a major general; died 1941.

10. Commissioned 1867; retired 1911 as a colonel; died 1914.

11. A small water craft used by the Coast Artillery Corps to lay mines to protect harbors and coastal areas from enemy ships.

12. Albert L. Mills; USMA 1879; retired 1916 as a brigadier general; died 1916.

13. Hugh Lennox Scott; USMA 1877; retired 1918 as a major general; was chief of staff of the army, 1914–1917; died 1934.

14. Thomas H. Barry; USMA 1877; retired 1919 as a major general; died 1919.

15. Frederick W. Sibley; USMA 1874; retired 1916 as a brigadier general; died 1918.

16. Fred Winchester Sladen; USMA 1890; retired 1931 as a major general; died 1945.

17. The young daughter of Captain John B. Christian (USMA 1896; retired 1922; died 1938), who was an assistant professor at West Point 1905–1909.

13. To Texas and Fort Myer

1. On March 8, 1911, President William Howard Taft ordered the concentration of thirty thousand troops, assembled as the Maneuver Division and commanded by Major General William H. Carter, at San

Antonio, TX, for ninety days. Although the stated reason for the gathering of many separate army units was a mobilization test and tactical maneuvers, Taft was concerned that revolutionary unrest in Mexico might spill across the border, as it later did. See Ronald G. Machoian, *William Harding Carter and the American Army* (Norman: University of Oklahoma Press, 2005), 231–37; and Clarence C. Clendenen, *Blood on the Border: The United States Army and the Mexican Irregulars* (New York: Macmillan, 1969), 144–50.

2. USMA 1905; retired 1946 as a major general; died 1966.

3. Colonel Joseph Garrard; USMA 1873; retired 1914 as a colonel; died 1924.

4. The troops at Fort Myer were the ceremonial units that supported military funerals at adjacent Arlington National Cemetery.

5. In 1907, President Teddy Roosevelt ordered officers to take an annual physical fitness test that included a ninety-mile horseback ride.

6. Frederick Dent Grant; USMA 1871; son of U. S. Grant; retired 1906 as a major general; died 1912.

7. Colonel Garrard.

8. 1872–1936; was the attorney general of the United States 1919–1921.

9. Founded in 1910 by the industrialist Andrew Carnegie with a grant of $10 million, the former secretary of war Elihu Root as president, and the charge from Carnegie "to hasten the abolition of international war, the foulest blot on our civilization," the Carnegie Endowment for International Peace still exists in 2010.

10. Rogers commanded the unit throughout its federal service in 1916, but not when it was refederalized in 1917.

11. Barry, previously superintendent at West Point when Summerall was on the faculty, at this time was the commanding general of the Eastern Department.

12. Eugene Gifford Grace (1876–1960) was president of Bethlehem Steel.

13. Captain Gerald E. Griffin; was veterinarian for the Third Field Artillery 1912–1914; retired 1922 as a colonel; died 1935.

14. The War Department

1. Summerall testified about appropriations for militia training camps before the House Military Affairs Committee on December 8, 1914.

2. Regular army officers assigned to monitor the organization and training of state militia units.

3. On Connecticut Ave. in northwest Washington, DC.

4. Early in World War II, it was utilized as an artillery training range.

5. Robert M. Danford; USMA 1904; retired 1942 as a major general and the chief of field artillery (1938–1942); died 1974.

6. Charles G. Treat; USMA 1882; retired 1922 as a brigadier general; promoted 1930 to major general on the retired list; died 1941.

15. The World War

1. An independent War Department mission, headed by Colonel Chauncey B. Baker (USMA 1886), visited France, conferred with American Expeditionary Forces (AEF) staff, and provided recommendations to the War Department on organization and policy. Summerall was the artillery specialist on the Baker Mission, but his recommendations for overwhelming artillery support to break the deadlock in the trenches was not followed, a continuing point of contention between him and the staff at AEF General Headquarters (GHQ).

2. Major General Tasker H. Bliss; USMA 1875; was later (1917–1919) the U.S. representative at the Supreme War Council; retired 1919; died 1930; was promoted posthumously to general, the rank he held as chief of staff during World War I.

3. Brigadier General William A. Mann; USMA 1875; retired 1918 as a brigadier general; died 1930.

4. Summerall's time line is a bit muddled here as Pershing and his party departed New York May 28, 1917, aboard the SS *Baltic*, arriving in Liverpool June 8.

5. John G. Quekemeyer; USMA 1906; was one of Pershing's aides for most of the war; died 1926 as a colonel.

6. Commander (Acting Captain) Bertram F. Hayes; Royal Naval Reserve.

7. The HMT *Olympic* was a British passenger liner and the sister ship to the *Titanic*, requisitioned during the war as a troop transport.

8. Arthur Balfour (1848–1930), British Conservative politician and foreign secretary. In April and May 1917, he headed a mission to the United States to discuss coordination of the war effort.

9. The *Olympic* departed Halifax June 2, 1917, arriving in Liverpool June 9.

10. Robert H. Dunlap; commissioned 1878 in the U.S. Marine Corps (USMC); died 1931 as a brigadier general; was not actually part of the Baker Mission but an artillery expert detailed to AEF GHQ.

11. William A. Lassiter; USMA 1889; retired 1931 as a major general; died 1954.

12. Lady Nancy Witcher Langhoren Astor (1859–1925); a Virginian

by birth and the wife of Viscount William Waldorf Astor, a friend and confidant of the U.S. ambassador, Walter Hines Page.

13. Lord George Nathaniel Curzon (1859–1925), a Tory member of Prime Minister David Lord George's cabinet.

14. Earl Edward George Stanley Derby (1865–1948); was British war minister 1916–1918.

15. Lord Frederick Roberts ("Bobs") was a British general, one of the heroes of the late-Victorian army, who commanded British forces in South Africa during the Boer War; he died in 1914. Consequently, Summerall most likely refers here to Field Marshal Sir William Robertson (1860–1933), at this time chief of the Imperial General Staff.

16. A broad leather waist belt with a diagonal strap over the shoulder. It was not approved in U.S. Army uniform regulations and eventually became a point of contention between the War Department and the AEF.

17. According to the final report of the Baker Mission, the visit was to the headquarters of the Second British Army, commanded by General Herbert Plumer (1857–1932).

18. Nelson E. Margetts; commissioned 1901; died 1932 as a colonel.

19. Major J. R. C. F. Xavier Reille; awarded the U.S. Distinguished Service Medal in 1920 for service with a French military mission to the United States.

20. General Henri Gouraud (1867–1946).

21. The Baker Mission report and related documents are reproduced in *The United States Army in the World War, 1917–1919*, 17 vols. (Washington, DC: Department of the Army, Historical Division, 1948), 1:55–115. For Summerall's differences with AEF GHQ over field artillery, see ibid., 110–15.

22. The Paris home of Ogden Mills (1884–1937), who was a wealthy American with many political and business connections with the French, a close friend of Major Robert Bacon on Pershing's staff, prominent in the Republican Party, and President Hoover's undersecretary and secretary of the treasury 1927–1933.

23. Alfred Charles William Harmsworth Northcliff (1865–1922) was a British newspaper proprietor who owned *The Times* and the *Daily Mail*.

24. 1869–1952; the assistant secretary of war and director of munitions.

25. Tasker Bliss.

16. Over There

1. The son of Summerall's mentor (same name; in the Fifth Artillery), who was killed in China in 1900. Reilly the son (1881–1963);

USMA 1905; resigned 1914 as a first lieutenant; recommissioned during the war; returned to civilian life after the Armistice; was the first president of the Reserve Officer Association 1922–1923; retired as a brigadier general, Officer Reserve Corps, 1948; one of Summerall's best friends.

2. 1877–1947; commissioned in federal service August 5, 1917, as a colonel; discharged May 24, 1919; a banker in Indianapolis.

3. 1876–1955; commissioned in federal service August 5, 1917, as a colonel; discharged July 14, 1919; served on active duty as a major general and the chief of the Militia Bureau 1931–1934; was a four-term mayor of Minneapolis.

4. USNA 1892; retired 1936 as a rear admiral; died 1948.

5. Percy W. Foote; USNA 1901; retired 1930 as a rear admiral; died 1961.

6. Edouard V. M. Isaacs; USNA 1915; retired 1927 as a lieutenant commander; died 1991.

7. About 9:00 A.M., May 31, 1918, three torpedoes fired by U-90 hit the ship, which sank within twenty minutes; 26 of the 715 men aboard went down with the ship.

8. Alban B. Butler (1891–?) of Washington, DC, commissioned August 1917, was Summerall's aide throughout the war. The two men remained friends afterward when Butler became a successful businessman working in the oil industry in Oklahoma. Butler's diary, from December 23, 1917, to October 12, 1918 (the period Summerall commanded the First Field Artillery Brigade and the First Division), is reproduced in vol. 11 of *World War Records of the 1st Division, AEF* (Washington, DC: Society of the First Division, 1928–1930). I have been unable to locate the original copy of Butler's diary, the published version of which does not cover Summerall as Fifth Corps commander, but there are copies of half a dozen informative letters from Butler to his family (dating from late October to mid-November 1918) in box 23, folder 1st Division—Commendations, Summerall Papers. The First Division Museum at Cantigny, Wheaton, IL, has a version of the diary, but essentially only what is reproduced in *World War Records of the 1st Division*.

9. There is nothing in Captain Stirling's November 27, 1917, voyage report about this. In fact, the report indicates an "uneventful trip," but rough weather throughout, before landing in St. Nazaire November 2. It includes an enclosed thank-you note to Stirling from Summerall. See Subject File, 1910–1927, "OS–President Lincoln," RG 45, NA.

10. Summerall's observations seem based on his dealings with the French during the spring and summer of 1917, when they were still recovering from the disastrous Nivelle Offensive. More nuanced historical accounts, including the state of French morale, are related in Robert B. Bruce, *A Fraternity of Arms: America and France in the Great War* (Law-

rence: University Press of Kansas, 2003); and Robert A. Doughty, *Pyrrhic Victory: French Strategy and Operations in the Great War* (Cambridge, MA: Harvard University Press, 2005).

11. Most of the AEF's heavy weapons and equipment came from overseas French and British sources.

12. Noble Brandon Judah (1884–1929), the Forty-second Division intelligence officer, was ambassador to Cuba 1927–1929.

13. According to the Sixty-seventh Field Artillery Brigade war diary (RG 120, NA), this occurred November 30, 1917.

14. Albert d'Amade (1856–1941).

15. December 1, 1917, according to the brigade war diary (RG 120, NA).

16. Summerall's service during the war is most closely connected to the First Division, the history of which is well told in James Scott Wheeler, *The Big Red One: America's Legendary 1st Infantry Division from World War I to Desert Storm* (Lawrence: University Press of Kansas, 2007).

17. A French barrack hut that widened toward the ground level to provide extra floor space.

18. This is a paraphrase from *Field Service Regulations, U.S. Army, 1914 (Corrected to July 31, 1918)* (Washington, DC: U.S. Government Printing Office, 1918), 74.

19. The diary of Alban Butler, Summeral's aide, indicates that from January 17 to January 25, 1918, Summerall was suffering from a cold that affected his eyes and resulted in doctors confining him to quarters. See *World War Records of the 1st Division*, vol. 11.

20. After the war, Steamer, who on occasion corresponded with Summerall, worked with the police department of the Southern Railway, in Washington, DC.

21. Long tubes of explosives used to clear a path through wire entanglements.

22. During this period, the division was under operational control of French Thirty-second Corps.

23. Georges Clemenceau (1841–1929), from November 1917 the French war minister and leader of the government.

24. Marie Eugene Debeney (1864–1943); French corps and army commander.

25. The brigade headquarters monthly "returns and officers' rosters" (RG 407, NA) do not reveal the identity of this officer.

26. The Austro-German Caporetto Offensive against the Italian army during October–November 1917.

27. Summerall's declaration might have been true, but neurosis or "shell shock" definitely was a problem in the AEF. See Thomas W.

Salmon and Norman Fenton, *Neuropsychiatry in the American Expedition-ary Forces,* vol. 9, *The Medical Department of the United States Army in the World War* (Washington, DC: U.S. Government Printing Office, 1929).

28. Colonel Malin Craig; USMA 1898; chief of staff of the army 1935–1939; retired 1939 as a general; died 1945.

29. Major General Clarence R. Edwards; USMA 1883; retired 1922 as a major general; died 1931; was relieved from command of the Twenty-sixth Division on October 25, 1918.

30. General Joseph Alfred Micheler (1861–1931) commanded the Fifth French Army at this time.

31. General Charles M. E. Mangin (1866–1925) commanded the Tenth French Army at this time.

32. 1870–1970; chairman of the War Industries Board 1918–1919; a financier and New York state executive who during World War II was an adviser to the Roosevelt administration on economic mobilization.

33. 1873–1931; a member of the War Shipping Board; was a lawyer and J. P. Morgan Co. executive.

34. One of the best tactical analyses of any AEF battle is Allan R. Millett, "Cantigny," in *America's First Battles, 1776–1965,* ed. Charles E. Heller and William A. Stoft (Lawrence: University Press of Kansas, 1986), 149–85.

35. The professional journal of the U.S. Infantry Association was a vehicle for discussion of organization, tactics, and equipment.

36. Colonel Bertram T. Clayton; USMA 1886; killed in action May 30, 1918.

37. A creosol and resin soap compound used as a disinfectant and deodorant.

38. According to A. W. Butler's July 13 diary entry, Summerall was at AEF headquarters, Chaumont, for discussions about the changing command arrangements, but, according to *Order of Battle of the U.S. Land Forces in the World War: AEF Divisions* (Washington, DC: U.S. Government Printing Office, 1931), he did not assume command of the division until July 15.

17. Soissons—the Decisive Battle of the War

1. A nearly definitive analytic account is Charles V. Johnson II and Rolfe L. Hilman Jr., *Soissons, 1918* (College Station: Texas A&M University Press, 1999).

2. Actually, the First Moroccan Division received relief on the night of July 20, having fought for three days, while the Second American Division was relieved on July 19, after two days.

3. See n. 31, chapter 16, above.

4. According to Johnson and Hilman, *Soissons,* 181 n. 12, the visit actually took place on the third day, July 20.

5. Colonel Campbell King; commissioned regular army 1898; retired 1933 as a major general; died 1953.

6. Lieutenant Colonel Clark R. Elliott; commissioned regular army 1901; died July 21, 1918, at Soissons; was second in command of the Twenty-sixth Infantry Regiment.

7. Brigadier General Beaumont Buck; USMA 1885; retired 1924 as a colonel; promoted 1930 to major general on the retired list; died 1950; commanded the Second Infantry Brigade at Soissons.

8. Colonel Hamilton A. Smith; USMA 1893; killed in action on July 22, 1918, at Soissons while commanding the Twenty-sixth Infantry Regiment.

9. For more on AEF formations, see War Department General Staff, Historical Branch, *A Study in Battle Formation* (Washington, DC: U.S. Government Printing Office, 1920).

10. The June–July 1917 Baker Mission.

11. Elliott.

12. Colonel Frank Parker; USMA 1894; retired 1936 as a major general; died 1947; commanded the Eighteenth Infantry Regiment at Soissons.

13. Conrad S. Babcock; USMA 1898; retired 1937 as a colonel; promoted 1941 to brigadier general on the retired list; died 1950; commanded the Twenty-eighth Infantry at Soissons.

14. General de Brigade Albert Joseph Marie Daugan (1866–1952).

15. Campbell King.

16. Frank Parker.

17. Major Redmond C. Stewart; from Baltimore; direct commission as a major July 17, 1918; discharged May 1, 1919.

18. Major General Pierre Emile Berdoulat (1861–1930), French Twentieth Corps.

19. Major General Hamilton Lyster Reed (1869–1931); retired 1928.

20. Each AEF division had an organic sanitary train—essentially a battalion-sized unit of six hundred men whose principal role was casualty evacuation.

21. Captain Barnwell R. Legge; Citadel 1911; commissioned regular army 1916; retired 1948 as a brigadier general; died 1949; was commanding a company in the Twenty-sixth Infantry at this time.

22. These were field artillery pieces mounted on motor truck chassis to enhance mobility.

18. Recovery

1. As commander of the First Division between Soissons and St. Mihiel, Summerall held a number of conferences with brigade and

regimental commanders and staff on organization, tactics, and discipline. Surviving notes of those meetings seem to emphasize artillery support for infantry advances, communications to front and to rear, and liaison with adjacent units. See the several memos dated August 25 and 26, 1919, in box 2, folder July–October 1918, Summerall Papers.

2. 1870–1946; was a New York City banker and New York State comptroller 1909–1911; appointed as the Red Cross representative with the First Division.

19. The St. Mihiel Salient

1. One of the best AEF battle analyses is Mark E. Grotelueschen's *The AEF Way of War: The American Army and Combat in World War I* (New York: Cambridge University Press, 2007), a comparative study of four divisions. For a good description of the First Division at St. Mihiel, see ibid., 105–25.

2. USMA 1881; retired 1921 as a major general; died 1927; clashed later with Summerall over the Sedan incident.

3. Brigadier General Harold B. Fiske; USMA 1897; retired 1935 as a major general; died 1960. The AEF G-5 (assistant chief of staff for training) was observing First Division operations during the St. Mihiel offensive. GHQ representatives frequently visited AEF divisions while in the line.

20. The Second Phase of the Meuse-Argonne

1. An excellent tactical account is Rexford C. Cochrane, "The 1st Division in the Meuse-Argonne: 1–12 October 1918," in *Gas Warfare in World War I: Study No. 2* (Washington, DC: U.S. Army Chemical Corps Historical Studies, 1957). For a discussion of the First Division's grueling fight in the second phase of the Meuse-Argonne, see Grotelueschen, *The AEF Way of War*, 125–41.

2. The effort of the First Division was very significant for the First Army attack to oust German defenders from the Argonne Forest. See Edward M. Coffman, *The War to End All Wars: The American Military Experience in World War I* (New York: Oxford University Press, 1968), 323–25.

3. Colonel Hugh A. Drum; commissioned 1898; retired 1943 as a lieutenant general; died 1951.

4. John N. Greely; commissioned 1908; retired 1943 as a major general; died 1965.

5. Hugh Drum.

6. This document is reproduced as "Summaries of Intelligence, 1917–18," in vol. 4 of *World War Records of the 1st Division.*

7. See Frederick Palmer, *Our Greatest Battle* (New York: Dodd, Mead, 1919).

8. Major General George H. Cameron; USMA 1883; retired 1924 as a colonel; promoted 1930 to major general on the retired list; died 1944.

9. Actually, Cameron was USMA 1883, while Pershing graduated in 1886.

10. Brigadier General Wilson B. Burtt; USMA 1899; retired 1938 as a brigadier general; promoted 1942 to major general on the retired list; died 1957.

11. Major General William G. Haan; USMA 1889; retired 1920 as a major general; died 1920.

12. Brigadier General Edwin B. Winans; USMA 1891; retired 1933 as a major general; died 1947.

13. Brigadier General Frank R. McCoy; USMA 1897; retired 1937 as a major general; died 1954.

14. Even consulting a variety of sources, it has not been possible to identify this officer. Six of Summerall's classmates were commanding regiments at this time, but none were in the Thirty-second Division.

15. For further information about this action, see Robert H. Ferrell, *The Question of MacArthur's Reputation: Côte de Châtillon, October 14–16, 1919* (Columbia: University of Missouri Press, 2008).

16. Major General Charles T. Menoher; USMA 1886; retired 1926 as a major general; died 1930.

17. Brigadier General Michael J. Lenihan; USMA 1897; retired 1929 as a brigadier general; died 1958.

18. Colonel Harry D. Mitchell; commissioned regular army 1899; retired 1938 as a colonel; died 1945.

19. William J. Donovan (1883–1959); was a New York lawyer and prewar militia officer; won the Medal of Honor for his action on October 14–15, 1918; served as director of the Office of Strategic Services in World War II.

20. Mitchell.

21. Lieutenant Colonel Charles Dravo; commissioned regular army 1904; retired 1942 as a colonel; died 1958.

22. Colonel Benson W. Hough (1875–1935); attended Ohio State University; commissioned as a colonel when his national guard unit was federalized in July 1917 and commanded the 166th Infantry; was a justice on the Ohio Supreme Court 1920–1923 and a U.S. district court judge 1923–1925.

23. The story of the Forty-second Division in the Meuse-Argonne is

well told in Robert H. Ferrell, *America's Deadliest Battle: Meuse-Argonne, 1918* (Lawrence: University Press of Kansas, 2007), 102–9; and James J. Cooke, *The Rainbow Division in the Great War, 1917–1919* (Westport, CT: Praeger, 1994), 165–208.

24. Menoher.

25. Lenihan.

26. Colonel Henry J. Reilly.

27. Mitchell.

28. Dravo.

29. Lieutenant General Hunter Liggett; USMA 1879; retired 1921 as a major general; promoted 1930 to lieutenant general on the retired list; died 1935. Liggett and Lenihan (USMA 1887) were not classmates.

30. Summerall is generally acknowledged as among the more effective tactical commanders in the AEF, particularly in his use of artillery support. He certainly was seen as being aggressive and an activist. Among other sources, Coffman's *The War to End All Wars*, Ferrell's *America's Deadliest Battle*, Millett's *The General*, and Wheeler's *The Big Red One* address his effectiveness as a commander.

31. George H. English, *History of the 89th Division, U.S.A.* (Denver: Smith-Brooks, 1920), 167–68.

32. Major General William M. Wright (see n. 2, chap. 4, above).

33. Haan.

34. Captain Charles H. Gerhardt; USMA April 1917; retired 1952 as a major general; died 1976; commanded the Twenty-ninth Division in World War II.

35. Menoher.

36. Major General John A. Lejeune, USMC, commanded the division; USNA 1888; was USMC commandant 1920–1929; retired 1919 as a major general; died 1942. Brigadier General Wendell C. Neville, USMC, commanded the Fourth Brigade; USNA 1890; died 1930 while serving as major general commandant. Colonel Robert O. VanHorn commanded the Third Brigade; commissioned regular army 1898; retired 1940 as a brigadier general; died 1941.

37. Drum.

38. Parker took over the First Division when Summerall was promoted to the Fifth Corps command.

39. At this time, Paul Maistre (1858–1922) commanded the French Groupe d'Armées du Centre.

40. For discussions of American employment of artillery in this attack, see Mark E. Grotelueschen, *Doctrine under Trial: American Artillery Employment in World War I* (Westport, CT: Greenwood, 2001), chap. 6, and *The AEF Way of War*, 266–78.

41. Lejeune.

42. Another reference to Summerall's participation in the Baker Mission discussions of organization and tactics in June 1917.

43. 1837–1921; was a soldier, railroad executive, and diplomat; as U.S. ambassador to France (1897–1905) initiated a search that led to the discovery, identification, and return of the remains of John Paul Jones to the United States.

44. This document is reproduced in *The United States Army in the World War, 1917–1919*, 9:385.

45. Frank Parker.

46. Menoher.

47. Brigadier General George G. Gatley; USMA 1890; died 1930 on active duty as a colonel.

48. Henry J. Reilly, *Americans All: The Rainbow at War* (Columbus, OH: F. J. Heer, 1936).

49. For more on the controversy surrounding the attack on Sedan, see Coffman, *The War to End All Wars*, 348–54; and Smythe, "A.E.F. Snafu at Sedan."

50. Colonel Robert P. Johnston, 314th Engineers; USMA 1893; died 1923 as a colonel.

51. These losses were not light. In fact, Summerall himself acknowledged that they were "excessive." See Grotelueschen, *Doctrine under Trial*, 127.

52. 1830–1882; USMA 1850. Another interpretation of his relief after Five Forks (April 1, 1865) was that Sheridan did not believe that Warren was sufficiently aggressive in pursuing Lee's retreating army.

53. Major General Harry C. Hale; USMA 1883; retired 1925 as a major general; died 1946.

54. This took place February 1, 1919, at Prauthoy, France. See Signal Corps photo SC-52159, RG 111, NA.

55. Major General George B. Duncan; USMA 1886; retired 1922 as a major general; died 1950.

56. Alvin C. York (1886–1964); earned the Medal of Honor for action on October 8, 1918, while fighting in the Meuse-Argonne. See David D. Lee, *Sergeant York: An American Hero* (Lexington: University Press of Kentucky, 1985).

57. For the rest of his life, Summerall often was obsessed with Sedan and the blemish it left on his otherwise outstanding tactical record. See the exchange of letters in April 1939 he had with Frank Parker (commanding the First Division during the Sedan incident), box 11, folder Miscellaneous, 1939–1940, Summerall Papers.

58. James G. Harbord; commissioned regular army 1891; retired 1922 as a major general; died 1947.

59. The AEF created a succession of post-Armistice boards to study

organization, tactics, weapons, and equipment. Summerall here likely refers to the Westervelt Board on artillery.

60. Unable to identify to whom Summerall refers here. Elsie Janis was the most famous of the entertainers who went to France during the war, but she lived much longer (1890–1956) than indicated here.

61. Brigadier General William K. Naylor; commissioned regular army 1898; retired 1938 as a brigadier general; died 1942.

62. The Thirty-fifth Division, originally composed of Kansas and Missouri national guardsmen. See Robert H. Ferrell, *Collapse at Meuse-Argonne: The Failure of the Missouri-Kansas Division* (Columbia: University of Missouri Press, 2004). In March 1919, the AEF chief of staff, Major General James W. McAndrew, asked Summerall to assess and attempt to calm the situation in two national guard divisions, the Twenty-sixth and the Thirty-fifth, where relations between the state national guard officers and AEF GHQ had become troublesome. See the Summerall and McAndrew exchanges in box 2, folder March–April 1919, Summerall Papers.

63. Joel Bennett Clark (1890–1945), U.S. senator from Missouri 1933–1945.

64. Dwight Davis (1879–1945); secretary of war 1925–1929.

65. Harry S. Truman (1884–1972); president of the United States 1945–1953.

66. Major General Clarence R. Edwards; USMA 1883; retired 1923 as a major general; died 1931.

67. Colonel Duncan K. Major; USMA 1899; retired 1940 as a brigadier general; died 1947.

68. Colonel Richard K. Hale; Harvard University 1902; discharged from federal service in 1919 as a colonel; was a brigadier general in the Massachusetts National Guard in 1923.

69. Mrs. George A. McKinlock, Lake Forest, IL. Her son, Second Lieutenant George A. McKinlock, had been killed in action on July 21, 1918, at Soissons while serving with the Third Machine-Gun Battalion, First Division. See McKinlock's burial file, RG 92, NA.

21. The Adriatic and Peace

1. The site of the French foreign ministry.

2. On July 8, 1919, the American Commission to Negotiate Peace authorized the Inter-Allied Commission of Inquiry on Fiume to examine the circumstances surrounding clashes between Italian and French troops at Fiume on the Adriatic, which in previous days had resulted in exchanges of gunfire and loss of life. Fiume is currently the Croatian port of Rijeka. In addition to Summerall, the members of the commis-

sion were Major General Sir H. E. Watts of Great Britain, Major General Stanislas Naulin of France, and Lieutenant General Mario Nicolis di Robilant of Italy. Communications from the Inter-Allied Commission of Inquiry on Fiume to the American Commission to Negotiate Peace, including the final report, are reproduced in vols. 7 and 11 of *Papers Relating to the Foreign Relations of the United States: The Paris Peace Conference, 1919* (Washington, DC: U.S. Government Printing Office, 1945, 1946).

3. Dundas L. Bartolucci; later served as the Italian assistant naval attaché in Washington, where he died in 1920.

4. Rear Admiral Philip Andrews; USNA 1886; retired 1930 as a rear admiral; died 1935.

5. By the Treaty of London, April 16, 1915, in exchange for political, colonial, and strategic concessions, Italy joined France and England in the war against Germany and Austria-Hungary. Although Summerall claims that this was a secret document, the general terms of the treaty were immediately known to, among others, the U.S. ambassador in Italy, Thomas Nelson Page. See U.S. Department of State, *Papers Relating to the Foreign Relations of the United States, 1915: Supplement: The World War* (Washington, DC: U.S. Government Printing Office, 1928), 31–39.

6. Summerall exaggerates official American unconcern and disinterest, as is clear from the communications and report reproduced in vols. 7 and 11 of U.S. Department of State, *Papers Relating to the Foreign Relations of the United States: The Paris Peace, Conference, 1919.*

7. On September 12, 1919, Gabriele d' Annunizo (1863–1938), an Italian writer, nationalist, and war hero, seized Fiume with a band of volunteers. The Italian government disavowed his actions.

8. Vittorio Emanuele Orlando (1860–1952); Italian prime minister 1917–1919; succeeded in June 1919 by Francesco Nitti (1869–1953).

9. Victor Emmanuel III (1869–1947); king of Italy 1900–1946; died in exile in Egypt.

10. Peter Augustus Jay (1877–1933); Harvard University 1900; a State Department foreign service officer 1902–1927.

22. Home and the First Division

1. A former German passenger liner (ex-*Vaterland*) seized by the United States in 1917 and used as a troopship. It departed Brest September 1, 1919, and arrived in New York September 8.

2. Sous-Lieutenant Edouard Gouin had served with Summerall throughout the war and earned the Distinguished Service Medal (War Department General Order 126, 1919) for his services.

3. It is unclear to whom he refers, as ship passenger files for the

Leviathan indicate that Bartolucci was accompanied on this trip by Seaman Augustine Zolozi. See RG 92, NA.

4. 1872–1959. Diplomat, lawyer, newspaper publisher, he graduated from the University of Pennsylvania in 1891 and served for twenty years in the State Department but also saw military service in 1898. During World War I, he was the liaison officer between Pershing's headquarters and the British War Office.

5. Root (1845–1937), the former secretary of war (1899–1904), was also a Republican senator from New York (1909–1915). At this time, he was the chairman of the Carnegie Endowment for International Peace.

6. Brigadier General George A. Wingate; commanded the Fifty-second Field Artillery Brigade, Twenty-seventh Division, during the war; commissioned August 5, 1917, as a colonel in the New York National Guard; promoted April 12, 1918, to brigadier general in the U.S. Army; discharged from federal service March 31, 1919.

7. 1871–1942; special assistant in the State Department Division of Western European Affairs 1918–1919; was also the long-serving president of the College of Charleston (1897–1942).

8. Appointed second lieutenant of field artillery on November 29, 1916; served with the First Division throughout the war; resigned his commission November 12, 1920.

9. Captain Harry B. Vaughn Jr. (1888–1964); was adjutant of the First Engineer Regiment at this time; commissioned regular army 1920; retired 1946 as a major general, at which time he was serving as a military aide to President Truman.

10. Charles G. Dawes (1869–1951). A banker and best friend of John Pershing, he was the AEF general purchasing agent in Europe during World War I. He served as vice president of the United States during the Coolidge administration.

11. 1843–1926; was a son of the president; had been secretary of war 1881–1885; served as minister to England during the Benjamin Harrison administration.

12. J. Ogden Armour (1863–1927); Chicago businessman and heir to the meatpacking firm.

13. Presumably the railway sleeping car family.

14. John Singer Sargent (1856–1925); American artist.

15. Abbott Lawrence Lowell (1856–1943); was president of Harvard 1909–1933.

16. John H. Sherburne; Massachusetts National Guard; entered federal service August 5, 1917, as a colonel; was an artillery officer in the Twenty-sixth Division during World War I; promoted June 26, 1918, to brigadier general; left federal service April 27, 1919.

17. Herbert Putnam (1861–1955); librarian of Congress 1899–1939.

18. 1859–1934.

19. 1850–1931.

20. Clarence O. Sherrill; USMA 1901; at this time was superintendent of public buildings and parks; resigned from the army 1926 as a colonel to become the city manager of Cincinnati; died 1959.

21. Frank B. Brandegee (1864–1924); Republican representative 1902–1905 and senator 1905–1924.

22. USMA 1924; commanded a field artillery battalion in World War II; retired 1954 as a colonel; died 1979.

23. 1873–1921; Democratic senator from Alabama 1911–1921.

24. An inflammation or infection of the mastoid bone, which sits behind the ear. The mastoid contains air cells that drain the middle ear.

23. Camp Dix, New Jersey

1. Society of the First Division, *History of the First Division during the World War, 1917–1919* (Philadelphia: John C. Winston, 1922).

2. Unable to provide further identification.

3. The anniversary of the First Division's first offensive in 1918.

4. *World War Records, First Division, A.E.F., Regular,* 31 vols. (Washington, DC: Society of the First Division, 1928–1930).

24. Hawaii

1. George J. Forster; commissioned November 27, 1917; earned Distinguished Service Cross for action with the Twenty-sixth Infantry commanding a section of thirty-seven-millimeter guns in the Meuse-Argonne October 4–13, 1918 (War Department General Order 14, 1928); retired 1951 as a colonel; died 1979.

2. Florian D. Giles; commissioned 1917 Officer Reserve Corps; discharged 1922; earned the Distinguished Service Cross with the Twenty-sixth Infantry at Montdidier May 27, 1918 (War Department General Order 39, 1920).

3. USMA 1884; commanded Hoboken Port of Embarkation during the war; retired 1925 as a major general; died 1940.

4. USMA 1892; retired 1922 as a colonel; died 1925.

5. Between January and May 1922, Summerall conducted thirty-three inspections of units, barracks, hospitals, and depots in the Hawaiian Department. His notes of these visits stress "discipline, training, morale, fortitude, community, enemies, objective—contentment." They also stated a tenet of his leadership—"all impulses come from the top." See box 4, folders 1921–1923 and 1921–1924, Summerall Papers.

6. Refers here to the 1921–1922 Washington conference between the

United States, Great Britain, France, Italy, and Japan, which placed limits on the naval building of the participants.

7. USMA 1904; was the commanding general of army ground forces in World War II when he was killed in action in 1944 as a lieutenant general.

8. Charles H. White; USMA 1907; retired 1946 as a major general; died 1971.

9. USNA 1905; retired 1947 as a fleet admiral; died 1966.

10. Summerall's war planning in Hawaii was forward thinking, as described in Brian Linn, *Guardians of Empire: The U.S. Army and the Pacific, 1902–1940* (Chapel Hill: University of North Carolina Press, 1997), 195–97. But Summerall's account here is written with some hindsight, from a perspective after December 7, 1941. The problem of the defense of Oahu was immensely complicated, with many competing interests. See Edward S. Miller, *War Plan Orange: The U.S. Strategy to Defeat Japan* (Annapolis, MD: Naval Institute Press, 1991).

11. Colonel Irving J. Carr; commissioned regular army 1898; retired 1934 as a major general; died 1963.

12. Vice Admiral Sir Frederick Laurence Field (1871–1945); commanded the Royal Navy Special Service Squadron, which included the HMS *Hood*.

13. Rear Admiral Edward Simpson Jr.; USNA 1880; retired 1924; died 1930.

14. Rear Admiral John D. McDonald; USNA 1884; retired 1927 as a vice admiral; died 1951.

15. Commander John V. Babcock; USNA 1901; retired 1936 as a captain; died 1955.

16. 1879–1936; commissioned regular army 1901, signal corps; resigned 1925 in the wake of his court-martial conviction. At this time the assistant chief of the air corps, Mitchell undertook a trip inspecting military aviation in the Pacific and Asia December 1923–July 1924. His report was particularly critical of the situation in Hawaii. For accounts of his Hawaii visit, see Alfred F. Hurley, *Billy Mitchell: Crusader for Airpower* (Bloomington: Indiana University Press, 1975), 86–89; and James J. Cooke, *Billy Mitchell* (Boulder, CO: Lynne Rienner, 2002), 151–67.

17. Lorrin A. Thurston (1858–1931) was a prominent Honolulu lawyer and businessman and the publisher of the *Pacific Commercial Advertiser.*

18. This facility remains in existence today as the Kilauea Military Camp, a Joint Services recreation center.

19. First lieutenant Ralph D. Sproul; commissioned regular army 1918; retired on disability 1928 as a first lieutenant.

20. Prince Jonah Kuhio Kalaiana-ole Pi'ikoi (1871–1922). He was not the last prince, but according to one source (www.hawaiianency-clopedia.com): "He was given the last state funeral held for a Hawaiian *ali'i* (royalty)." He was active for a long time in Hawaiian politics, including organizing the Republican Party in the territory and serving as the second nonvoting representative in the U.S. House of Representatives.

21. The earthquake at Yokohama on September 1, 1923, killed 140,000 over two days of quake and fire.

22. The Red Cross representative with the First Division in the AEF.

23. 1854–1932; Democrat from Mississippi; served in the House of Representatives 1893–1909 and in the Senate 1910–1923.

24. Probably Charles C. Kearns (1869–1931); Republican from Batavia, OH; served in the House of Representatives 1915–1931.

25. Return to the States

1. On August 20, 1920, the War Department created nine corps area commands embracing the continental area of the United States. The corps area commanders, usually major generals, were responsible for overseeing administration, training, and tactical control of units within their area, to include the reserve components as well as the regular army.

2. 1874–1939; merchant, philanthropist, and civic leader from Chicago.

3. Yale University 1892; commissioned regular army 1897; was the chief of staff of the Second Division and then the commanding general of the Third Division in the AEF; retired 1934 as a major general; died 1948.

4. 1850–1924.

5. There were strikes and work stoppages during the war, but the AFL's support for the war effort kept these to a minimum. Gompers and other labor leaders also had representation on most of the U.S. government war boards, including the Council of National Defense. See "Labor in America," in Cochran, ed., *Concise Dictionary of American History*, 530.

6. The Eighth Corps Area chief of staff was Colonel John F. Preston; USMA 1894; retired 1936 as a major general; died 1950.

7. Summerall previously has indicated having a hand in relieving five officers during the war—Conrad Babcock, Beaumont Buck, Robert P. Johnston, Michael Lenihan, and Harry D. Mitchell. In 1924–1925, Buck was retired from the army but living at Fort Sam Houston.

8. Unable to identify this officer.

26. From Texas to New York

1. John Huston Findley (1863–1940); philanthropist, journalist, educator; was an editor at the *New York Times* as well as the president of the City College of New York.

2. General Gerardo Muchado (1871–1939), the Liberal Party candidate, was president of Cuba 1925–1933.

3. 1846–1927; president of U.S. Steel Corp.

4. Sir Esme Howard (1863–1939); British ambassador to the United States 1924–1930.

5. Appears here to be referring to Fort Hamilton, Brooklyn, NY, the headquarters of First Division at this time.

6. Later Fort Drum, near Watertown, NY.

7. Patrick E. Crowley (1864–1953); president of the New York Central Railroad.

8. Officer classification boards met during the early 1920s to determine the fitness of individuals to continue on active duty. Selection boards determined the fitness of individual officers for promotion.

9. Summerall initially was one of twelve general officers selected to sit on the court-martial convened to judge Mitchell on the charges of conduct prejudicial to "good order and . . . discipline" and conduct that had brought "discredit to the military service," these charges stemming from Mitchell's criticism of his superiors' (military and civilian) policies on airpower. Early in the court-martial, which lasted from October to December 1925, Mitchell's defense challenged Summerall and one other general on the court for "prejudice and bias," and they were dismissed. But Summerall later testified for the prosecution against Mitchell. For Summerall's testimony, see Douglas Waller, *A Question of Loyalty: Billy Mitchell and the Court Martial That Gripped the Nation* (New York: Harper Collins, 2004), 308–9.

10. Alfred E. Smith (1873–1934) was the unsuccessful Democratic candidate for president in 1928 and a four-term governor of New York (1919–1920, 1923–1928).

11. 1856–1943; president of Harvard University 1900–1933.

12. See Donald Kington, *Forgotten Summers: The Story of the Citizens' Military Training Camps* (San Francisco: Two Decades, 1995).

13. 1870–1951; succeeded Samuel Gompers as president of the AFL.

14. Edward J. Bowes (1874–1946); the "patron saint" of American amateur performers; the initiator of "The American Amateur Hour"; was an army intelligence officer in World War I.

15. 1881–1934.

16. The principal Union Civil War veterans' organization.

17. 1904–?; the daughter of Colonel Russell P. Reeder Sr. (1872–1943),

who was commissioned in 1898 and retired in 1936 as a colonel; also the sister of the army athletic legend and author "Red" Reeder (1902–1998, USMA 1926), who commanded the Twelfth Infantry, Fourth Division, in Normandy, where he earned the Distinguished Service Cross.

18. Brigadier General Robert E. Callan; USMA 1896; retired 1936 as a major general; died 1936.

27. Chief of Staff

1. Butler was Summerall's wartime aide.

2. According to his "Diaries, Notes, and Engagement Books," while chief of staff Summerall had as many as twenty luncheons, dinners, receptions, etc. per month. In addition, between March 1927 and May 1930, he made forty-two inspection trips outside Washington, DC. Although many were to army installations along the Eastern seaboard, several were to the Far West and the West Coast; travel was usually by railroad. See box 1, Summerall Papers.

3. Major General Fox Conner; USMA 1898; was deputy chief of staff March 9, 1926–April 30, 1927; retired 1938 as a major general; died 1951.

4. Hanford McNider (1889–1966); served as the assistant secretary of war February 20, 1926–December 15, 1932; was a farmer and banker born in Iowa; during World War I earned the Distinguished Service Cross with oak leaf cluster for service with the Ninth Infantry, Second Division; was also the national commander of the American Legion.

5. Major General Briant H. Wells; USMA 1894; retired 1935 as a major general; died 1949.

6. Dwight F. Davis (1879–1945); Harvard University 1901; was the assistant secretary of war March 5, 1923–October 13, 1925, and the secretary of war October 14, 1925–November 18, 1929; served as the governor general of the Philippines 1929–1932.

7. Theodore Roosevelt, *The Rough Riders* (New York: Charles Scribner, 1899).

8. During the period he was chief of staff, Summerall testified twenty times before congressional committees, five times for Senate committees and fifteen for House committees. Most frequently, his testimony concerned the annual War Department appropriations bill, but other issues included supplemental funding for army construction projects, especially family housing; the enlisted reserve; the promotion system; the publication of world war records; and aerial coast defense. For a listing of and specific citations to these hearings, see Congressional Information Service (CIS), *Congressional Committee Hearings Index*, pt. 3, *69th–73d Congress, 1925–1934* (Washington, DC: CIS, 1984), 496–97.

9. McNider.

10. Each major country considered a potential threat had its own color, e.g., red for Britain, black for Germany, yellow for China. *American War Plans, 1919–1941: Plans for War against the British Empire and Japan*, ed. Stephen T. Ross (New York: Garland, 1992), includes copies of a number of the plans.

11. See Miller, *War Plan Orange*.

12. On February 8 and 9, 1904, Japanese destroyers launching torpedoes, not submarines, did most of the damage to the Russian fleet at Port Arthur. See Christopher Martin, *The Russo-Japanese War* (New York: Abelard-Schuman, 1967), 39–58.

13. Most American strategists of the time recognized that Corregidor could not be held indefinitely, and it was not in 1942. War Plan Orange–3, in effect in the Philippines in 1941, and previous iterations actually required U.S. and Filipino forces in the face of a Japanese attack on the Philippines to concentrate in central Luzon, gradually withdrawing south to the Bataan Peninsula. The objective was to hold the entrance to Manila Bay, deny its use to the Japanese navy, and prepare to receive a relief expedition. See Louis Morton, *The Fall of the Philippines: The United States Army in World War II* (Washington, DC: Department of the Army, 1953), 61–64.

14. In 1927, the Joint Board members included Summerall, Briant Wells (deputy chief of staff), Colonel Stanley Embick (chief of the War Plans Division), Admiral Edward W. Eberle (chief of naval operations), Rear Admiral Thomas J. Senn (assistant chief of naval operations), and Rear Admiral Frank H. Schofield (director of the Naval War Plans Division).

15. F. Trubee Davison (1898–1976); Yale University 1918; naval aviator in World War I when he earned the Navy Cross; assistant secretary of war for air February 20, 1926–December 15, 1932.

16. Fechet was chief of the air corps December 20, 1927–December 19, 1931; commissioned regular army 1900 from the volunteers; retired 1931 as a major general; died 1948.

17. The principal mechanism for the changes was the Air Corps Act of 1926, which provided funds and a legal basis for air corps expansion and some autonomy, including creating the position of the assistant secretary of war for air.

18. Richard D. Newman; USMA 1908; retired 1939 as a colonel; died 1939.

19. Calvin Coolidge.

20. Major General B. Franklin Cheatham; commissioned regular army 1899; retired 1930 as a major general; died 1944.

21. John M. Morin (1868–1942); Republican congressman from Pennsylvania; was chairman in both the Sixty-ninth and the Seventieth Congresses.

22. Frank B. Kellogg (1856–1937); lawyer and Republican party official; also served New York in the U.S. Senate 1912–1923.

23. James W. Wadsworth Jr. (1877–1952); Republican from New York; was chairman of the Senate Committee on Military Affairs through the Sixty-ninth Congress (to March 4, 1927). Wadsworth was defeated for reelection and was succeeded as chairman by David Aiken Reed (1880–1953), a Republican from Pennsylvania.

24. Brigadier General Briant H. Wells was the G-4 during the early period that Summerall was chief of staff (March 9, 1926–April 30, 1927); retired 1935 as a major general; died 1949.

25. Major General Clarence C. Williams; USMA 1894; was chief of ordnance 1918–1930; retired 1930 as a major general; died 1958.

26. For information about the 1924 Philippine Scout "revolt" and concern over other internal security threats in the archipelago, see Linn, *Guardians of Empire*, 148–49, 157–59.

27. Referring to the Washington conference on naval limitations.

28. Major General Preston Brown.

29. Bok Tower is the centerpiece of Bok Gardens, created in 1921 as a bird sanctuary at Lake Wales, FL, by Edward W. Bok, the editor of the *Ladies Home Journal.*

30. Summerall was only the sixth officer in history to attain that rank in the U.S. Army, the others being Washington, Grant, Sherman, Sheridan, and Pershing. During World War I, Tasker Bliss and Peyton C. March had held emergency rank as full generals.

31. Was chief of field artillery December 20, 1927–February 15, 1930; commissioned U.S. Volunteers 1898; commissioned regular army 1901; retired 1930 as a major general; died 1939.

32. Major General Clarence C. Williams.

33. Colonel Tracy C. Dickson; USMA 1892; retired 1932 as a brigadier general; died 1936.

34. 1865–1944. Christie's problems with the Ordnance Department are described in George F. Hofmann, "A Yankee Inventor and the Military Establishment: The Christie Tank Controversy," *Military Affairs,* February 1975, 12–18.

35. Colonel Daniel Van Voorhis; commissioned U.S. Volunteers 1898; commissioned regular army 1900; retired 1942 as a lieutenant general; died 1955.

36. It is unclear to whom Summerall actually refers here because the assistant chief of staff for operations and training (G-3 Division), War Department General Staff, had officers in both its Mobilization Branch and its Training Branch concerned with the national guard and the reserve.

37. The most complete account of these and subsequent administra-

tions' economy measures and their impact on the army is John W. Killi-grew, "The Impact of the Great Depression on the Army, 1929 to 1936" (Ph.D. diss., Indiana University, 1960). Edward M. Coffman (*The Regulars: The American Army, 1898–1941* [Cambridge, MA: Harvard University Press, 2004], chaps. 7–8) also discusses these issues, especially the impact on individual soldiers and officers.

38. *Overseas casuals* in this context refers to individual soldiers temporarily unassigned to any unit while in transit to or from overseas assignments, mostly in the Pacific insular possessions.

39. 1866–1929; U.S. representative from Iowa 1909–1921; secretary of war March 5–November 18, 1929; died in office.

40. 1883–1963; assistant secretary of war March 15–December 8, 1929; secretary of war December 9, 1929–March 4, 1933. During World War II, President Roosevelt sent the Republican Hurley on several sensitive diplomatic missions.

41. 1860–1936; also served Kansas in the U.S. House and Senate.

42. Colonel David L. Stone; USMA 1898; retired 1940 as a major general; died 1959; was Hurley's military assistant but does not appear to have ended his career "in revolt."

43. This was the 1948 Hoover Commission, which reported on federal government organization.

44. Hoover's interior secretary was Ray L. Wilbur (1878–1945).

45. November 18, 1929.

46. Unable to further identify.

47. 1890–1973; Princeton University 1914; served with the U.S. mission to Austria 1919–1920; was the U.S. ambassador to Austria 1930–1933.

48. Frederick H. Payne (1876–1954); banker and businessman; assistant secretary of war December 20, 1929–April 5, 1933.

49. Summerall's "final report" is Chief of Staff, *Annual Report of the War Department: Fiscal Year Ended June 30, 1930* (Washington, DC: U.S. Government Printing Office, 1930).

50. Despite stories, rumors, and legends to the contrary, there was no rash of suicides among students at the Command and General Staff School during this period. Between 1919 and 1940, only two students committed suicide, one in June 1924 and the second in December 1924, both before Summerall was chief of staff. See Timothy K. Nenninger, "Casualties at Leavenworth: A Research Problem" (paper presented December 12, 2009, at University of Wisconsin History Department and Wisconsin Veterans Museum symposium).

51. These interwar boards reviewed individual officers' records in an effort to eliminate the unfit, thereby helping reduce the large "hump" or "bulge" of those commissioned during World War I.

28. The Wanderer's Return

1. Duncan Upshaw Fletcher (1859–1936); Democrat from Florida; U.S. Senate 1903–1936.

2. John P. Thomas to Charles P. Summerall, December 19, 1930, box 9, folder December 1930–December 1931, Summerall Papers.

3. Oliver James Bond (1865–1933); South Carolina Military Academy 1886.

29. The Citadel

1. John P. Thomas (1873–1949); Citadel 1893; was chairman of the Board of Visitors 1925–1949.

2. Lieutenant Colonel John W. Lang; USMA 1907; retired 1946 as a brigadier general; died 1967.

3. Williams and McCormick, previously mentioned in the World War I chapters, were colleagues of Summerall's from the First Division.

4. Jesse H. Metcalf (1860–1942); Republican senator from Rhode Island 1924–1937.

5. Aliss A. V. Jones married Frank Geary January 30, 1911.

6. Dennis P. Quinlan; enlisted 1899 volunteer service; commissioned regular army 1901; retired 1934 as a brigadier general; died 1936.

7. Summerall's report to the Board of Visitors on this incident, dated April 17, 1933, is in box 9, folder The Citadel, 1931–1933, Summerall Papers.

8. Frank Hamilton Bowles (1907–1975) was the Columbia University registrar and director of admissions 1937–1968.

9. 1871–1955; was a Democratic congressman from Tennessee 1907–1931; secretary of state 1933–1944.

10. Clarence M. McMurray (1886–1972); South Carolina Military College 1910; commissioned regular army 1910; served on active duty through World War II; retired 1948 as a colonel.

11. Florence C. Gaston (1896–?); the wife of Colonel Jesse Gaston, one time commandant of cadets; was a hostess at the school 1938–1955.

12. The secretary of war established the Army Specialized Training Program in late 1942 to ensure the procurement of technically and professionally trained men for the broad range of specialties required by the army. When manpower shortages, particularly infantry replacements, became manifest in early 1944, the program was terminated.

13. A high church Episcopalian congregation, following the Anglo-Catholic tradition.

14. Burnet Rhett Maybank (1899–1954); mayor of Charleston 1931–1938; governor of South Carolina 1939–1941; Democratic senator from South Carolina 1942–1954.

15. James Ripley Westmoreland (1876–1960); Citadel 1900.

16. Charles Francis Myers (1892–1957); Citadel 1914; retired 1957.

17. 1885–1968.

Bibliography

On Summerall

Addington, Larry H. "Charles Pelot Summerall." In *Dictionary of American Military Biography*, ed. Roger J. Spiller, 1077–80. Westport, CT: Greenwood, 1984.

Anger, Charles L. "Charles Pelot Summerall." In *Dictionary of American Biography: Supplement 5*, 668–69. New York: Scribner's, 1977.

Mossman, B. C., and M. W. Stark. "Former Army Chief of Staff, General Charles P. Summerall, Combined Services Full Honor Funeral, 14–17 May 1955." In *The Last Salute: Civil and Military Funerals, 1921–1969*, 87–88. Washington, DC: Department of the Army, 1991.

Nenninger, Timothy K. "Charles P. Summerall as Chief of Staff of the Army, 1926–1930." 1986. Typescript.

Nichols, W. Gary. "The General as President: Charles P. Summerall and Mark W. Clark as Presidents of the Citadel." *South Carolina Historical Magazine* 95, no. 4 (October 1994): 314–35.

———. "General Charles P. Summerall: The Training, Command, and Education of the Citizen-Soldier." In *Unknown Soldiers: The American Expeditionary Forces in Memory and Remembrance*, ed. Mark A. Snell, 144–64. Kent, OH: Kent State University Press, 2008.

———. "General Charles P. Summerall and the Dilemmas of the Peacetime Army, 1926–1930." N.d. Typescript in editor's possession.

Pratt, Fletcher. "Charles P. Summerall." In *Eleven Generals: Studies in American Command*, 250–54. New York: William Sloan, 1949.

Summerall, Charles P., Jr. "Charles Pelot Summerall." *Assembly*, January 1956, 57–58.

On the U.S. Army

Coffman, Edward M. *The War to End All Wars: The American Military Experience in World War I*. New York: Oxford University Press, 1968.

———. *The Old Army: A Portrait of the American Army in Peacetime, 1784–1898*. New York: Oxford University Press, 1986.

———. *The Regulars: The American Army, 1898–1941*. Cambridge, MA: Harvard University Press, 2004.

Dawson, Joseph G. *The Late 19th Century U.S. Army, 1865–1898: A Research Guide*. Westport, CT: Greenwood, 1990.

Ferrell, Robert H. *America's Deadliest Battle: Meuse-Argonne, 1918*. Lawrence: University Press of Kansas, 2007.

Fletcher, Marvin. *The Peacetime Army, 1900–1941: A Research Guide*. Westport, CT: Greenwood, 1988.

Griffith, Robert K., Jr. *"Men Wanted for the U.S. Army": America's Experience with an All-Volunteer Force between the World Wars*. Westport, CT: Greenwood, 1982.

Grotelueschen, Mark E. *Doctrine under Trial: American Artillery Employment in World War I*. Westport, CT: Greenwood, 2001.

———. *The AEF Way of War: The American Army and Combat in World War I*. New York: Cambridge University Press, 2007.

Hallas, James H. *Squandered Victory: The American First Army at St. Mihiel*. Westport, CT: Praeger, 1995.

Johnson, Douglas V., II, and Ralph L. Hillman Jr. *Soissons, 1918*. College Station: Texas A&M University Press, 1999.

Killigrew, John W. "The Impact of the Great Depression on the Army, 1929–1936." Ph.D. diss., Indiana University, 1960.

Linn, Brian McAllister. *Guardians of Empire: The U.S. Army and the Pacific, 1902–1940*. Chapel Hill: University of North Carolina Press, 1997.

———. *The Philippine War, 1899–1902*. Lawrence: University Press of Kansas, 2000.

Millett, Allan R. *The General: Robert L. Bullard and Officership in the United States Army, 1881–1925*. Westport, CT: Greenwood, 1975.

———. "Cantigny, 28–31 May 1918." In *America's First Battles, 1776–1965*, ed. Charles E. Heller and William A. Stofft, 149–85. Lawrence: University Press of Kansas, 1986.

———. "Over Where? The AEF and the American Strategy for Victory, 1917–1918." In *Against All Enemies: Interpretations of American Military History from Colonial Times to the Present*, ed. Kenneth J. Hagan and William R. Roberts, 233–56. Westport, CT: Greenwood, 1986.

Nenninger, Timothy K. "The Army Enters the Twentieth Century, 1904–1917." In *Against All Enemies: Interpretations of American Military History from Colonial Times to the Present*, ed. Kenneth J. Hagan and William R. Roberts, 219–34. Westport, CT: Greenwood, 1986.

———. "American Military Effectiveness in the First World War." In *Military Effectiveness* (2 vols.), ed. Allan R. Millett and Williamson Murray, 1:116–56. Boston: Allen & Unwin, 1988.

Roberts, William R. "Reform and Revitalization, 1890–1903." In *Against All Enemies: Interpretations of American Military History from Colonial Times to the Present*, ed. Kenneth J. Hagan and William R. Roberts, 197–218. Westport, CT: Greenwood, 1986.

Smythe, Donald. *Pershing: General of the Armies*. Bloomington: Indiana University Press, 1986.

Spector, Ronald. "The Military Effectiveness of the US Armed Forces, 1919–1939." In *Military Effectiveness* (2 vols.), ed. Allan R. Millett and Williamson Murray, 2:70–97. Boston: Allen & Unwin, 1988.

Weigley, Russell F. *History of the United States Army.* New York: Macmillan, 1967.

———. "The Interwar Army, 1919–1941." In *Against All Enemies: Interpretations of American Military History from Colonial Times to the Present,* ed. Kenneth J. Hagan and William R. Roberts, 257–77. Westport, CT: Greenwood, 1986.

Wheeler, James Scott. *The Big Red One: America's Legendary 1st Infantry Division from World War I to Desert Storm.* Lawrence: University Press of Kansas, 2007.

Wilson, John R. M. "Herbert Hoover and the Armed Forces: A Study in Presidential Attitudes and Policy." Ph.D. diss., Northwestern University, 1971.

Index

Because of the nature of a memoir, the text often refers to individuals by title or rank, instead of by name; for example, the reader will not find Dwight F. Davis mentioned by name on page 161, but rather as "secretary of war." Military units are indexed as spelled; for example, *Fifth Corps* appears before *First Corps*.